LINKED ARMS

LINKED ARMS

A Rural Community
Resists Nuclear Waste

THOMAS V. PETERSON
photographs by STEVE MYERS

STATE UNIVERSITY OF NEW YORK PRESS

Published by
STATE UNIVERSITY OF NEW YORK PRESS
ALBANY

© 2002 State University of New York
All rights reserved
Printed in the United States of America

For information, address
State University of New York Press,
90 State Street, Suite 700, Albany NY 12207

Production, book, and cover design, Laurie Searl
Marketing, Michael Campochiaro

Library of Congress Cataloging-in-Publication Data
Peterson, Thomas V., 1943–
    Linked arms : a rural community resists nuclear waste / Thomas V. Peterson ;
photographs by Steve Myers.
        p. cm.
    Includes bibliographical references and index.
    ISBN 0–7914–5131–3 (alk. paper) — ISBN 0–7914–5132–1 (pbk : alk. paper)
    1. Low level radioactive waste disposal facilities—Location—New York
(State)—Allegany County.   2. Low level radioactive waste disposal facilities—New York
(State)—Allegany County—Public opinion.   3. Public opinion—New York
(State)—Allegany County.   4. Environmental protection—New York (State)—Allegany
County—Citizen participation.   5. Civil disobedience—New York (State)—Allegany
County.   6. Nonviolence—New York (State)—Allegany County.   I. Title.

TD812.3.N7 P48 2001
363.72'8953'0974784—dc21   2001049332

10   9   8   7   6   5   4   3   2   1

For Robert McAfee Brown, mentor and friend, whose example showed me that a scholar often has the ethical duty to become an activist,

and

for Carol Burdick, colleague, author, and friend whose belief in my writing abilities gave me confidence to write this nonfiction account in a novelistic style,

and

for the people of Allegany County whose many personal sacrifices preserved the environment and provided a unique model of nonviolent resistance for people everywhere.

Contents

Maps and Illustrations

PHOTOGRAPHS

MAPS

Preface

IN A RURAL COUNTY IN western New York, seventy miles south of Rochester, ordinary men and women not only kept a major nuclear dump out of their area, but also provided a model of resistance for communities across the United States. Merchants, teachers, homemakers, farmers, and blue collar workers ignored potential jail terms and large fines to challenge the nuclear industry and the government. The events in this story transformed common folks into extraordinary individuals, many of whom temporarily gave up their personal lives for the welfare of their community.

When New York State's siting commission said that Allegany County would be a good place for low level nuclear waste, men and women asked questions. Some answers troubled them; others horrified them.

"What is low level nuclear waste?"

The siting commission said it was not too dangerous—just things such as booties and gloves that workers wore inside nuclear power plants; or needles and test tubes that doctors used to treat cancer patients; or petri dishes and irradiated animals used by researchers in colleges and universities.

That was only part of the story, Marvin Resnikoff, a prominent nuclear physicist, told Stuart Campbell, a founder of the resistance movement in Allegany County. "Low level" waste included almost everything except for spent fuel rods, the casing around the reactor's core, and tailings from uranium mining. It did include, however, the irradiated metal clamps that hold the plutonium fuel rods in place in the core of nuclear reactors. Some of the material was so "hot" that it would kill anyone exposed to it for more than a few minutes; some had such long half-lives that it would remain dangerous for hundreds of thousands of years.

This question led to many others. "How can you choose the best site if you don't know how you're going to store the waste?" "Will you incinerate waste at the facility?" "What is the half-life of the cement that will encase the nuclear waste?" "How dangerous is exposure to low doses of radiation?"

Questions exceeded answers. People began reading about the routine failures of storing nuclear waste throughout the United States. The technicians, they read, always wrapped themselves in the mantle of expertise, assuring concerned citizens that "professionals knew best." Dismissing public concern as naive and ignorant, the experts had, in fact, made disastrous mistakes—then lied about them. Was it reasonable to think that these New York State commissioners had better solutions for handling nuclear waste than had their colleagues at Hanford, West Valley, Fernauld, Oak Ridge, and Rocky Flats?

The past tragedies of nuclear storage and the callousness of the experts who had ignored people's safety and health shocked the folks in Allegany County. Many wondered, however, whether a county that had no economic or political clout could defy the Empire State. Others joined together to form a movement that would contest the methodology of the siting commission, the statutes of New York, the short-term economic motivations of the nuclear industry, and the policies of the federal government.

The citizens hired a lawyer who challenged the congressional law and persuaded the governor to file a lawsuit against the federal government that eventually went all the way to the Supreme Court. They enlisted a preeminent nuclear physicist who convinced the governor that the classification of nuclear waste was bogus. They shattered the glib assurances of the siting commission. Most importantly, not content to question authority, they defied it. Five times over a twelve-month period they linked arms, stood in the bitter cold, and defeated the siting commission through civil disobedience.

Hundreds of people, who had never before broken the law, refused to let the siting commission with its multimillion dollar budget get onto the land. They refused to move when scores of state troopers ordered them to open up the roads. They defied a judge who threatened to fine them thousands of dollars and toss them into jail for six months.

These folks were optimists. They were also naive, believing they could persuade federal authorities to re-think their failed strategies for classifying and storing nuclear waste. But they were not lazy. These citizens sent out emissaries to other communities where authorities were trying to build nuclear dumps—in Nebraska, Connecticut, Michigan, Illinois, North Carolina, California—telling the story of Allegany County's defiance. "Resist!" they said. "Don't cooperate with those who want to put a nuclear waste dump in your community. Follow our example. Engage in civil disobedience! Join our lawsuit against the federal government!"

Allegany County's story, grounded in optimism and forged in struggle, is the story of American democracy, rooted in the Declaration of Independence, which provided a mythological framework for defiance of authority. Civil disobedience runs deep in American culture. Thoreau refused to pay taxes that would finance an unjust war; suffragettes voted illegally and went to jail to promote women's rights; northerners hid runaway slaves; civil rights protesters rode integrated buses into the deep South. Allegany County built upon these acts of rebellion. For this reason, if for no other, their story deserves to be told, for it is the story of the American character itself.

This story of resistance also raises profound questions that Americans need to address. What happens to democratic principles when corporations seeking short-term profits decide issues of life and death? How should we balance the interests of the many and the rights of the few? When is it morally right to defy the state through nonviolent resistance?

The Allegany County story also raises serious questions about how our nuclear past will affect our future. To what extent are we willing to pay the real costs of storing nuclear waste? Is there even a solution? To what extent will we sweep the problem under the rug by building dumps in poor communities—an African American community in the deep South, an Indian reservation in the West, a farming district in the Midwest, an economically depressed area in the Appalachian Mountains?

Since this is a story of ordinary people's resistance to a perceived injustice, it is important to let the men and women speak for themselves. Dialogue is therefore critical to the story. I have reconstructed this history from videotaped conversations, personal diaries and journals, newspaper reports, and people's memories, including my own. I have formally interviewed nearly fifty people and transcribed their comments onto more than 1,500 pages of text, and I have spoken to scores of others while checking the facts in the story.

Writing history always includes editing. All historians decide that some meetings are more important than others, that some people are critical to the central story and others are marginal, that some motives essentially contributed to a person's action while others did not. Conversations are the same. Although I have tried to be true to the issues and thoughts that people expressed in those conversations and even true to their style of speaking, I have had to condense them and rework them for the printed page. In the vast majority of instances, I created the dialogues from quotations in newspaper articles and from videotapes. (See Sources for a more complete explanation about the historical accuracy of the dialogues.)

Almost all of the people quoted in this story are still living in my county. I have known some of them personally for a long time; I got to know others quite well while protesting alongside them and interviewing them. I hope that they will respond as Sheriff Larry Scholes did when I read him my account of a meeting that he had with three state troopers from Albany: "I can't swear every word is exact, but I have no corrections to make; I got goosebumps when you read it, because you sent me back there." When I once wrote a religious history of people in the Old South, I had to imagine that they would think I got it right. Here, as I've checked and rechecked many events, the protagonists have already helped me to make many corrections.

My biggest regret is eliminating so many people's activities and stories in creating this unified narrative. Many will believe that I have left out important details and events. Literally hundreds of people spent days organizing marches and rallies, writing and singing songs, planning concerts and holding bake sales, creating quilts that traveled around the nation telling Allegany County's story, making huge pots of food to feed frozen protesters at the sites, writing letters to the editor, calling their friends and neighbors. Perhaps Stuart Campbell, one of the founders of the nonviolent resistance movement, should have the final word here: "We wouldn't have won if there was one less person. Everyone was essential!"

I am indebted to many people for their help. First, the newspaper reporters. I have read hundreds of their stories in researching and writing this book. Their exceptional work and extraordinary accuracy provided the basis for this story. Foremost among them are Kathryn Ross of *The Wellsville Daily Reporter* and Joan Dickenson of *The Olean Times Herald*. Videotapes that Peer Bode, a video artist at Alfred University, graciously shared with me were also very helpful; he and his students took many hours of footage during the conflicts between protesters and police.

I also want to thank Corrine Bandera, who not only helped me to collect these newspaper reports, but also created a system for organizing them. I also used newspaper stories and reports gathered by Pam Lakin that are now part of the special collections at Herrick Memorial Library at Alfred University. Ann Hoffman spent considerable time helping me use a computer to prepare the first map in the book.

From time to time, I called many individuals in the county to check details in the story. I must thank them collectively; they know who they are. I am especially indebted to the people who allowed me to tape and transcribe their interviews. They include the principal characters in the story and many others. I have put an asterisk by their names in the index of names, so I will not repeat the list here. I thank Dan Sass for giving me technical information about the geology of the county, Kathryn Rabuzzi for advice on writing creative nonfiction, and Alan Littell for helping me prepare portions of the manuscript for initial inquiries to publishers.

I owe a tremendous debt to Carol Burdick, Sally Campbell, Stuart Campbell, and Gary Ostrower, friends and colleagues who provided me with assistance in writing this book throughout all of its stages. They encouraged me to undertake the project and gave me confidence and advice along the way. They read several versions of the manuscript and helped to fine tune both its content and its style. The book has been vastly improved by their help. Lou Ruprecht and Ward Churchill also read the manuscript critically and suggested strategies for publication. In addition to them, I appreciate the efforts of Eric Somer and Megan Staffel Marks in helping to find a publisher.

I thank Craig Prophet for his work on creating four of the five maps for the book and am especially grateful that Steve Myers agreed to share his artistic photographs for use in the book.

Prologue

THE SHERIFF OF ALLEGANY COUNTY leaned back in his chair and considered his chances for reelection. Three months earlier he had resolved to put thoughts about his future on the back burner and focus on the difficult task that lay ahead. He had, in fact, convinced himself that his career in law enforcement was over. Only a couple of weeks earlier he had told the district attorney at lunch that no one would elect him dogcatcher after this whole affair was over. "What I'll have to do in the next few months will make me the most unpopular guy in the county."

Larry Scholes knew that in normal times a sheriff can assure his reelection by enforcing the law impartially, by preserving the peace, and by maintaining personal integrity. But these were not normal times. The citizens were unified against putting a nuclear dump in the county and many of them were willing to act outside the law, committing acts of nonviolent resistance. Far worse, Scholes feared that some people might start shooting.

"Sheriff" is the only elected law enforcement position in the United States, making sheriffs' departments unusually responsive to citizens' needs. This is the reason why many people bring problems to their county sheriffs more frequently than they do to state and local police. Allegany County in western New York was, however, an exception. Sparsely populated and relatively poor, it was one of only two counties in New York State that did not have its own road patrol. For all practical purposes the state police were the primary law enforcement agency in the county.

Sheriff Larry Scholes glanced up at his calendar, noting that it was Pearl Harbor Day. He was awaiting a delegation of state troopers to discuss a looming battle between protesters and the New York State siting commission, charged with finding a suitable place to build a nuclear waste dump. Scholes was pleased that the state police had requested the meeting. He was puzzled, though, that three high-level troopers from the superintendent's Albany office were driving nearly three hundred miles to conduct it. Usually, Lieutenant McCole, who was in charge of the district office, or Captain Browning from the regional office near Buffalo would ask for a meeting to coordinate efforts between the agencies. The state police must have recognized, Scholes surmised, that the sheriff's department in Allegany County could not handle large-scale protests, and they were preparing to take charge. The troopers from the superintendent's office in Albany, he figured, were coming to muscle him out of the way.

The sheriff had mixed feelings about this possibility. He approved of the state police taking primary responsibility for enforcing the state's efforts to site a nuclear dump, but he hoped he could still play a role in keeping things peaceful. During the last six months, he had carefully cultivated relations with leaders in the anti-dump movement in order to assure them that he and his men would remain calm. He had never identified with the movie image of the "gun totin'" sheriff who bullied hoodlums into submission. In fact, he rarely carried a gun and never wore a star on his chest; his uniform was a white shirt and tie, with a fleece-lined leather jacket for cold weather.

Larry Scholes had joined the sheriff's department on April Fool's Day 1973, when he was twenty-four years old; he was appointed undersheriff two years later. Now forty-one years old, he had just been elected to his third term as sheriff, a position he had held for nearly seven years. The sheriff's department had been his only career.

Now he faced an uncertain future. He knew he would have a serious challenger in the next election, requiring him to spend between three and four thousand dollars, nearly ten percent of his yearly salary. Far worse was his anxiety about providing financial stability for his family.

These personal concerns gave way to even deeper worries about maintaining the peace between irate citizens of the county and the technical team from the siting commission. Scholes had talked with Bill Timberlake, his undersheriff, about handling the protests and coordinating activities with the state police. They concluded that the most reasonable approach would be for the two of them to escort the technical team until it met resistance from the "citizens" (Larry always referred to the protesters as "citizens," even in private conversations). If the protesters insisted on blocking access to the land, then he would call in the state police after calming everyone down.

Scholes took special pride in his ability to defuse tense situations; his primary role, he believed, was to foster peace in the community. He selected deputies whose philosophy about law enforcement coincided with his own.

He had chosen Timberlake as his undersheriff seven years earlier not only because of his experience, but also because Timberlake genuinely liked people, was a sympathetic listener, and had common sense. His most recent job had been with the police in the village of Alfred, one of two college towns in the county. (The other was Houghton, home of a small Christian college affiliated with the Wesleyan Church.) Timberlake knew how to walk into a fraternity party late at night and quiet things down.

In all the ways that mattered Scholes and Timberlake worked well together in promoting community harmony. In other ways they were quite different. Scholes had done all of his law enforcement work locally; Timberlake had spent the first fifteen years of his career as a New York City cop before he and his wife decided to move to a healthier environment. Scholes was trim and meticulously dressed; Timberlake was a bit overweight and slightly rumpled. Newspaper reporters loved writing stories about the sheriff, because his statements were direct, factual, articulate, and grammatically correct. They sometimes wondered, however, whether they were getting the full story. When the undersheriff answered their questions, however, reporters knew that he was speaking unrehearsed, as thoughts continually bubbled to the surface of the conversation. But they were not always sure they had understood the import of what he was saying, and they had a much harder time separating conjecture from fact.

Just then, Timberlake poked his head into the sheriff's office. "Jim's arrived; I took him into the conference room and gave him some coffee."

"I'm glad he got here before the troopers arrived. I want to tell him what we've been thinking." The sheriff had invited district attorney Jim Euken to the meeting. Not only was he a friend, but this would be a good chance to make sure that law enforcement would be coordinated.

"Hello, Jim. I'm glad you came, since you'll be involved in deciding how to prosecute any of the citizens who are arrested. Three high ranking state troopers are coming from Albany to discuss things."

"From Albany? Isn't that rather unusual?"

"It sure is. But nothing like this has ever happened in the county before." The sheriff paused, and the D.A., glancing over at Timberlake, eased himself into a chair at the table. Timberlake raised an eyebrow, suggesting that there was more to this meeting than he had been told.

"I suspect," the sheriff continued, "that the state police have come to tell this country boy that they will be coordinating all activities and that I shouldn't get in the way."

"Well, that should be a relief to you. If the state police play the heavies, you're off the hook." Jim Euken smiled and added, "You might get elected again, after all."

Sheriff Caught in the Middle © Steve Myers 1990; all rights reserved.

Scholes also smiled, but right now he was more concerned about his public responsibilities than his personal future. "Maybe I can convince the state police that I should have some role in keeping everything from escalating into violence. After all, I know these people. Many are friends and neighbors, and it's still my job to keep everyone safe. I hope the police leave Charlie in charge. We work well together."

The sheriff was referring to Lieutenant Charles McCole who commanded the local unit of the state police. They got along well, thought alike, and could almost instinctively read each other's minds. Scholes paused and then said, "But I'm uneasy that Albany's getting so involved."

Timberlake continued Scholes's thought, "Whatever those troopers are here to talk about will be connected with the politics of siting the dump. And if they're coming out of the superintendent's office they'll be representing the governor's interests."

"We'll know soon enough," the sheriff said. To himself, he thought, "Mario Cuomo's henchmen."

Five troopers entered the conference room where the three county officials waited. Scholes and Timberlake warmly greeted the two with whom they'd worked closely. The local troopers introduced their superiors from Albany, two inspectors and a colonel, and the sheriff introduced the district attorney. They sat on opposite sides of the table.

The sheriff smiled. "Well, I guess it's no mystery that you're here to talk about law enforcement problems involving the nuclear dump."

They exchanged some pleasantries, then one of the inspectors got down to business. "The involvement of the sheriff's department is extremely important in local protests. You know the people in your county and are in a much better position than the state police to assist the siting commission."

Scholes stiffened in his chair. He realized the meeting was taking a very different turn than he had expected. The troopers weren't muscling him out of the way, after all; they seemed to be asking him to assume all the responsibility. The inspector continued, "If the state police were to come on the scene, then it's probable that the conflict would escalate."

The inspector paused and let his statement settle into the silence. Scholes found himself saying things that he was sure the troopers already knew. "Of course, we'll be happy to do whatever we can to reduce potential violence. We're certainly willing to accompany the exploration teams to the sites. But as you know, we won't be able to handle any serious difficulties that arise."

"We're sure you can handle the situation."

"You know that we don't have a road patrol. All my officers are exclusively assigned to guard duty in the county jail, except for the two who take care of civil matters, serving subpoenas and enforcing court orders. That only leaves Bill and me to deal with any protesters. As long as I've been in the sheriff's department, we've called upon the state police for help."

"Of course we'll work on criminal matters—burglaries, murders, and these sorts of things," responded the inspector, "but our success depends on the trust we've developed with the people here. If we get involved in local protests, we'll compromise that trust. Working with the technical team is a civil matter between the courts, the siting commission, and your office."

The sheriff, who always prided himself in remaining calm under fire, bristled. "If people block roads, that's not just a civil matter. The technical team will insist on getting onto the land, and there's no way that Bill and I can accomplish that. The citizens have said they'll do civil disobedience to keep them off the land. So I'll need to call for help."

"Well, since this matter is really one for the sheriff's department," the inspector replied, "you should call mutual aid for support."

Scholes and Timberlake had already considered the option of calling for assistance from the sheriff's departments in surrounding counties. Protocols were in place to invoke "mutual aid" in case of disasters that required a large number of police to respond to a sudden emergency. But here the citizens had already said they would break the law. Invoking mutual aid could easily cost the county well over one hundred thousand dollars after only a couple of encounters. And the sheriff would be solely responsible for siting a nuclear dump among his friends and neighbors.

"It's not clear to me that it's our responsibility to enforce the building of a nuclear dump for the state of New York, even if it's in our county. It seems to me that it's a state matter and therefore a job for the state police."

The inspector leaned forward. "Sheriff, you don't understand. We're telling you that you *will not* call the state police for any reason. You *will* invoke mutual aid. This is your problem and you're going to have to deal with it. The state police *will not* become involved."

Theatrical performance is part of all good law enforcement. Scholes, in fact, was a master in conveying authority or calm by slight changes in the way he looked, spoke, and acted. He used these techniques in public situations where something could easily get out of hand. Right now, however, the sheriff was offended that a fellow officer would bully him. Even worse, it was a bad performance. Lieutenant McCole seemed embarrassed and stared blankly at his hands, looking, Scholes thought, as if he wished he were a hundred miles away.

Scholes dropped his hands to his sides, stared back at the inspector, and lowered his voice. "My oath of office does not permit me to decide which laws of New York State I'll enforce and which laws I won't. Neither does the state police have the right to enforce certain laws while ignoring others. We've always called upon the state police to help enforce the laws, and we'll continue to do so."

The sheriff knew that anger was surfacing. It was no calculated act, but he had no inclination to stop. "Take this message back to Superintendent Constantine and Governor Cuomo from Sheriff Scholes—that's S-C-H-O-L-E-S. Sheriff Scholes says 'BULLSHIT!'"

1 The Struggle Begins

Low level nuclear waste is a misleading term. West Valley has low level waste that's so radioactive it has to be driven in shielded trucks. . . . They will tell you that in one hundred years all the radioactivity will be gone; that's not true.

—Carol Mongerson of West Valley Coalition

DECEMBER 21, 1988, ALMOND, N.Y.—

NEARLY A YEAR BEFORE Sheriff Scholes's meeting with the state police from Albany, Betsy Myers was baking Christmas cookies and thinking about holiday parties, while her husband, Steve, sat at the kitchen table, reading the *New York Times*. Suddenly he jumped up, startling Betsy, and moved to the counter. "Look at this! There's a good chance that New York State's going to put a nuclear waste dump only a few miles from our home." A map of the state disclosed thirty-two townships that had been identified as potential sites. Five were in Allegany County. "Geographical, geological and population concerns," the article stated, "removed much of the state from consideration as potential sites," and "regulations excluded from consideration Long Island, New York City and the Adirondack Park."

"Low level nuclear waste," according to the article, included "things like contaminated clothing and equipment . . . from hospitals, industry and utilities." Steve was reading the article aloud to Betsy, but stopped and interjected, "Like hell. That's only the tip of the iceberg."

Steve was a muscular man with short cropped hair and such a closely clipped beard that it appeared to many as though he simply hadn't shaved for a week or two. He had been involved in enough environmental movements to doubt that any nuclear waste would be innocuous. After graduating from Pratt Art Institute in New York City where he specialized in photography, he worked on one of the first ecological exhibits in the United States, entitled

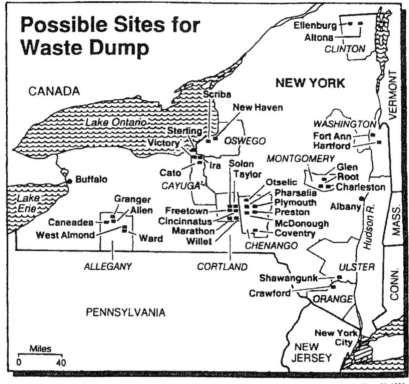

Possible Sites for Waste Dump

Ellenburg
Altona
CLINTON

CANADA

NEW YORK

Scriba
New Haven

Lake Ontario
Sterling
Victory
OSWEGO

WASHINGTON
Fort Ann
Hartford

VERMONT

Buffalo

Ira
Cato
CAYUGA

Solon
Taylor

MONTGOMERY

Glen
Root
Charleston

Otselic
Pharsalia
Plymouth
Preston

Lake
Erie

Granger
Allen
Freetown
Cincinnatus
Marathon
Willet
Ward

Caneadea
West Almond

McDonough
Coventry

Albany

Hudson R.

MASS.

CHENANGO

ALLEGANY

CORTLAND

ULSTER

CONN.

Shawangunk
Crawford
ORANGE

PENNSYLVANIA

New York
City

NEW
JERSEY

Miles
0   40

The New York Times / Dec. 21, 1988

Permission NYT Graphics

"Air and Water Pollution," at the Smithsonian Museum in 1967 and began an odyssey as a radical environmentalist. A year later he was on the staff of *New York Magazine* where he photographed stories that ranged from the Hell's Angels to the artistic community in SoHo. One particularly memorable piece was on pollution of the Hudson River. In addition to his work in photojournalism, Steve began to establish himself as a commercial photographer.

He and Betsy became sweethearts when both were students at Pratt. After graduation, they became pioneers in SoHo at a time when New York City zoning codes made it illegal to live there. Nevertheless, the inexpensive rents of abandoned industrial buildings and the huge spaces for lofts attracted struggling artists, and the authorities mostly ignored them.

Since the birth of their son, Matthew, a couple of years earlier, the Myerses had talked about moving to a rural area to raise their family. They had hesitated, because Steve feared his opportunities for artistic and commercial work would dwindle. Two muggings, one that sent Steve to the emergency room and the other that badly frightened Betsy, finally tipped the scales. They moved to Almond, a village on the eastern edge of Allegany County, not far from the Pennsylvania border.

Steve had earlier developed commercial relations with Dow Corning and Eastman Kodak, two companies situated in western New York within seventy miles of their new home. Betsy's family still lived in Almond, where her mother was highly respected as a local historian. Steve had been accompanying Betsy to the area since 1966. Having spent all of his youth in large American cities, he was fascinated by the county's rural life, which he began photographing.

Steve paced the kitchen and re-read the *New York Times* article out loud to Betsy, becoming more and more agitated. Looking again at the map showing the thirty-two targeted townships, he blurted, "I think it's going to be in Allegany County. The other places are either too close to New York City or too far from a major highway. They'll want to use the Southern Tier Expressway." He was referring to a four-lane highway (now designated Interstate 86) that bisects Allegany County, extending from Harriman near the Hudson to a corner of Pennsylvania where it intersects with Interstate 90. "Look at this map! Allegany County is the only targeted place in western New York. I know they'll try to put it here."

"You think so?" asked Betsy, a strikingly stylish woman in her forties.

"Sure. They'll try to find some rural community without much political clout. All the better if the place is near a major highway."

"What about the other places?" she asked, brushing her dark hair off her forehead with the back of her hand.

"There are a couple other possibilities, but Allegany County is just the sort of place that's most likely." Steve paused and added, "Remember when I traveled to that environmental conference in Bozeman, Montana?"

"Yeah, that's the conference you attended with Pete Emerson, wasn't it?"

Pete Emerson was then vice president of the Wilderness Society. He had been Betsy's classmate at Alfred-Almond Central School and had grown up on a nearby dairy farm that his parents owned and operated. His folks still supplemented their farm income by tapping sugar maple trees, boiling down the sap until it was forty times more concentrated. Impressed by Steve's work in environmental photography, Pete invited him to a conference in Bozeman, where he met other environmentalists and shared a cabin with Wisconsin's environmentalist senator, Gaylord Nelson.

"I heard enough about the shenanigans of the nuclear industry to know that Orazio's statement about no environmental hazard is just bullshit." Steve had again turned to the *New York Times* article, which quoted from Angelo Orazio, the head of the siting commission.

"Listen to this," he continued. "Orazio says that 'some localities might welcome the plant . . . because of potential jobs.' It's the same old pattern. That's the lure to fool people into accepting all kinds of hazardous shit!"

"We've got to alert people," Betsy said, little realizing how much their lives would be changed by this obvious statement.

Steve's single-mindedness and Betsy's dedication were largely responsible for the speed with which many groups in the southern and central parts of the county coalesced into a county-wide organization. In the final ten days of 1988, they telephoned everyone they knew and confronted people they saw in grocery stores and gas stations. Their words of alarm, like the seeds in one of Jesus' parables, were sown on both fertile and rocky ground, some quickly taking root, some withering on parched soil, some waiting for a good rain to get started. While people generally felt that a nuclear dump would threaten their lives, a few dismissed Steve as overzealous. In any case, most thought that battling the state would be futile. The Myerses realized they had a fight on their hands to overcome people's apathy.

Steve's language was peppered with images of callous officials in the nuclear industry, supported by corrupt politicians, who had knowingly destroyed communities across the United States for profit. Embedded in his conversations were snapshots of the ruination of idyllic life in the rural county. Most of all, a moral outrage sizzled beneath the surface of his words, a fury that occasionally left him speechless as he grappled to find appropriate language to express his horror about having a nuclear wasteland in his back yard.

Sandy Berry lived just a couple of blocks down the street from the Myerses and had been a good friend for several years. Steve called her shortly after he learned that the siting commission had targeted Allegany County. Her nine-year-old son, Garrett, was best friends with the Myerses' youngest child, Shep, also nine. Steve had recently taken photographs of her sculptural pieces for a New York Fine Arts Award application.

Sandy practically became part of the Myers family for the next few weeks and, in the beginning stages of the fight, would become Steve and Betsy's closest ally. A slight, animated young woman with an energy that matched her flaming red hair, Sandy's spirit was playful and relaxed, while Steve's energy was pointed and intense. When Steve's outrage caused him to lose perspective, Sandy would introduce lightheartedness into their discussions.

Over the next few weeks Sandy, a single parent, put her third-grade son on the school bus in the morning and walked down to the Myerses' house for coffee. The three of them would go to the post office to get the latest materials from protesters who were fighting a nuclear dump in Michigan, or from the West Valley Coalition, or, eventually, from their lawyer. As they walked to the post office, Sandy would spin dreams, casting Steve as a congressman and visualizing an environmental utopia. Other times she would speak of nightmares—people dying of cancer because of chemicals and radionuclides that greedy people were allowing into the environment. A few times her imagination even took her into a dark, mythological world where grotesque mutations occurred to her grandchildren because of her son's exposure to radiation.

In the week after Christmas, Betsy and Steve scheduled a meeting in Almond for the afternoon of January first. They never considered that people might want to watch football games that day. With no interest in sports and no television set, they were literally not plugged in to popular American culture. Remarkably, thirty people crowded into a room at the Historical Society's Hagadorn House for the hastily called meeting. Discussing the threat to the environment, Steve explained why Allegany County was the likely target for a nuclear dump.

Dressed in slacks and a blue shirt, he stood at the front of the room. "I know they're going to put it here," he said. "Our only chance is to get people together for massive protests. That'll be hard, because most people think you can't beat the state. It'll be tough, but if we can get enough people to stand up and confront the authorities, we can beat them. We've got to motivate the people in the county to resist. We've got to cover the county with posters opposing the dump."

Others at the meeting were cautious about getting too involved. Some privately wondered whether Steve was overly alarmist, even as they dutifully elected him as president of the tiny group. When no one else volunteered, Betsy, Sandy, and Betsy's brother also became officers. Steve was too focused on the impending confrontation to consider people's reluctance to take a more active role; he was gearing up for the battle of his life.

Not the first to organize—a group had formed in the northern part of the county two days earlier—the Almond group was particularly effective in coordinating the early activities in much of the county. Not only did Steve call people in other areas, encouraging them to become involved, he supplied them with information. Sandy Berry, with no scientific training, was their unlikely emissary, traveling in her 1958 Chevy to many early meetings where she passed out information and distributed pamphlets written by antinuclear activists in other parts of the country. The archetypal presence of wild energy in the guise of a woman with red hair moved many people to get involved.

❖

JANUARY 5, 1989 —

After putting her son on the school bus, Sandy Berry rushed down the street to the Myers house. She could hardly wait to tell them that nearly one hundred people had packed themselves into the tiny West Almond grange hall.

"Those farmers are really angry," she said, bursting into the Myerses' kitchen. "They're ready to fight. They wanted to know what to do. I told them we've got to blanket the county with our 'Bump the Dump' posters, and we need to raise money to pay for them. I also told them it was important to come protest when the siting commission comes here."

"Yeah, that's the first step to changing the apathy," Betsy said. "We've got to say loud and clear that we won't take it. We've got to keep the nukers from getting any foothold here at all."

Sandy shed her coat, helped herself to coffee and a doughnut, and sat down at the kitchen table. "Rich Kelley's got lots of farmers fired up in West Almond. His family's had a dairy farm there forever. He knows everyone and they seem to respect him."

"What's he like?" Steve asked.

"He's about my age, early thirties, strong stalky guy with a full, bushy beard. He's got lots of energy and his head's on straight. He graduated from Alfred University about fifteen years ago as a math major, but decided he'd rather farm his family's land than teach math and science in high school."

"We've got to invite him to attend our strategy sessions," Steve noted, helping himself to another cup of coffee.

"When I talked with him," Sandy said, "he wondered whether we were going to the political meeting tomorrow."

"What meeting?" Steve asked.

"He said that Hasper and Ostrower set up a meeting with county legislators and other big shots in the county." Sandy was referring to New York assemblyman John Hasper and his county liaison, Gary Ostrower, a professor of American history at Alfred University. "Rich Kelley's suspicious of politicians. He thinks Hasper's trying to take control and speak for the whole county."

"We can't let that happen." Steve sputtered, stood up, and paced across the kitchen. "The nuclear establishment will wrap the politicians in this county right around their little finger. They've probably already paid Hasper off.

"The state will promise to lower taxes and Hasper and the county legislators will sell out for blood money—thirty pieces of silver." As he talked,

Steve's eyes bulged and his face became red. "Don't they understand that this is war?"

"Calm down, Steve," Sandy interjected, putting her hand on his shoulder. "If you don't, your eyes are going to pop right out of your head and you'll have a heart attack."

Steve seemingly ignored her, but settled down. "We're going to attend this meeting! When and where is it?"

"I don't know," Sandy answered. "Rich didn't say. So many people were coming up to us after the meeting that we didn't talk long."

"Why don't we call Gary Ostrower and ask him?" suggested Betsy.

"Good idea," Steve responded. "Ostrower better know that he and Hasper won't get away with making deals behind our backs. Why don't you call him, Betsy?"

The phone rang in Ostrower's office at Alfred University. He answered the phone in his usual staccato style. "Ostrower."

"Hello, Gary. This is Betsy Myers. When and where is tomorrow's political meeting about the dump?"

"Two o'clock in the Legislative Chambers at the County Courthouse." Although Gary had done much of the organizing for the meeting by planning the agenda and suggesting people who should attend, Hasper had invited most of them. They had decided that presidents of the colleges, county legislators, prominent businesspeople, and scientists would more likely attend if the invitation came directly from the assemblyman. When he answered the phone, Gary had not recognized Betsy Myers's name and assumed that Hasper had invited her.

"You know that Steve and I have been organizing people to fight the nuke dump, and we'd like to come to your meeting," Betsy continued.

Seldom at a loss for words, Gary was momentarily startled into silence. They had wanted to keep the meeting small. Hasper and he had talked about getting the political and economic leaders of the county unified against the dump. Gary worried that hotheads spouting radical rhetoric would polarize the conservative, Republican county. Now he had visions of the strategy-planning session being invaded by large numbers of angry folks making unreasonable demands and antagonizing county leaders. He wanted to avoid this at all costs and blurted out the first thing that came to his mind. "You're welcome to come, but Steve is not."

Betsy was speechless. She couldn't think of anything to say, took a deep breath, and hung up the phone.

"What did he say?" Steve asked.

"He said I can come, but you can't."

"Bullshit, I can't!"

JANUARY 6, 1989—

Steve accompanied Betsy into the legislative chambers a few minutes before two o'clock.

John Hasper, a large man with graying hair, conveyed calmness and determination. Standing in front of the room, he set the tone. "You all know where I stand about siting a nuclear dump in the county. It would set back the progress we're making in attracting new industries. Frankly, I'm fed up to here with all the proposals to make us the dumping ground for other people's problems, whether it's ash dumps, nuclear waste, or garbage from New York City."

Politics was in Hasper's blood. He was born in Albany on election day, November 4, 1935. His father was a lawyer who worked for the state legislature's bill drafting commission for thirty-nine years, chairing it for the last sixteen. His mother was a legal secretary who worked for the commission. But rural life was also in his blood. He had attended school in Belfast, the geographical center of Allegany County, staying with his grandmother when the legislature was in session. He roamed the woods after school and became an avid hunter and fisherman.

As a state legislator, Hasper had fought hard to strengthen the economic life of the county that he deeply loved. A champion of the Southern Tier Expressway, he hoped that the major highway would attract new industries, particularly those compatible with the three colleges in the area: Alfred University, Houghton College, and Alfred State College. He was discouraged when the road only seemed to bring new proposals for building various kinds of garbage dumps.

"We've got to be smart about fighting this thing," he continued, punctuating his words with his hand. "We can't just start shooting off our mouths without knowing what we're talking about. The siting commission's coming here at the end of the month. We need to get as many people to come to that meeting as possible. We also need to organize speakers ahead of time so we roughly know who's going to talk about what."

Betsy Myers spoke up from the rear of the room. "Who'll decide who gets to speak?"

"We need to coordinate that," the assemblyman responded, "so that everyone doesn't say the same thing. I suggest we let the county administrator set it up. Why don't the leaders of the citizens' groups call his office?" Privately

Hasper shared Ostrower's concern that the people forming the protest movement were hotheads, who would alienate others. A conservative legislator, John saw many environmental protesters as "whackos," tree huggers who failed to consider economic issues along with their love for the land. He knew radicals would alienate people in the conservative county, and he vowed to keep them from controlling the county's fight against the dump.

Steve Myers, however, didn't hear the passion in Hasper's voice. He assumed that most politicians would say one thing, then turn around and do another. He imagined that the Republican legislator had major contributors who were part of the nuclear establishment. Steve leaned forward in his chair and stared intently at the legislator. He would keep the politicians from selling out the county. But he knew he'd have to act quickly to unify the various citizens' groups that were forming throughout the county.

This was as good a time as any, he thought, to establish his credentials. He rose and announced, "The citizens' groups will have an important meeting at the Almond fire hall next Friday at eight o'clock. Carol Mongerson of the West Valley Coalition will be there to explain the danger of nuclear dumps. We'll also be coordinating a strategy to confront the siting commission when they come on January 26."

Hasper didn't know whether Steve was one of those "whacko environmentalists," but he'd have to find out. For now, the assemblyman ignored the outburst and continued, "One more thing. I've asked Gary Ostrower to set up a technical committee to respond to the commission's report identifying Allegany County as a potential site. I used a similar technique in Livingston County when the Corps of Engineers wanted to dam the Genesee River. No one wanted to see the beautiful canyons in Letchworth State Park destroyed. We showed scientifically why it wouldn't be a good idea and were able to stop it."

Gary Ostrower, a short, thin man in his late forties, loved being at the center of political decision making. A liberal Republican, he seemed an unlikely spokesperson for the conservative legislator. Hasper was, however, impressed by Gary's energy and pragmatism; Gary deeply respected Hasper's integrity, even though he disagreed with him on many social issues such as abortion. "As most of you know," Gary said, "I work part time as John's liaison with the county. At his behest, I've started to put together a group of scientists and other experts to study the siting commission's report on the candidate areas. We're going to tear it apart and examine each piece of evidence."

Ostrower feared, however, that Hasper's earlier comments about his political commitment against the dump might scare off scientists committed to objective investigation. "I want to emphasize that this isn't going to be a hatchet job. We're going to examine all of the evidence and let the chips fall where they may. Our report will be credible only if we're scientifically objective."

What Ostrower did not say publicly was that he was putting an ethicist, a sociologist, and an economist on the technical committee to make sure that the report would contain serious objections to a nuclear dump, even if the geologists, biologists, and physicists concluded that the county was a suitable place for nuclear waste. Suspecting that the siting commission hadn't done its homework, however, Ostrower thought the scientists would find serious flaws in the report.

Steve Myers bristled when Ostrower talked about producing an objective report. Didn't he understand that this was war? Didn't the politicians understand that the road to hell was paved with scientific objectivity? "That's just what the siting commission wants," he muttered to himself, "people debating the fine points of science while the state is building a dump here."

Gary noticed Steve leaning forward in his seat, glaring at him. Irritated that Steve had crashed the meeting, Gary reiterated, "I've assured the scientists who are on the technical committee that they are expected to approach the data with complete objectivity. That's the only way that our report will have any credibility." Gary knew he was taking a risk, but felt that the odds were in the county's favor.

During the meeting a couple of geological experts offered a few observations about gas wells in the area and the possibility that the Clarendon-Linden fault extended into the county. Several citizens suggested other ways to thwart the siting commission. Rich Kelley proposed that the county legislature enact a law, requiring the state to do a baseline study of residents' health before they could build a nuclear dump. That would burden the state financially, he explained, and perhaps discourage them from choosing Allegany County. Hasper was enthusiastic. "I like it. That's the kind of idea we're looking for. We should have the county attorney look into this."

Near the end of the meeting Hasper introduced John Hunter, president of Alfred State College. Steve Myers became especially agitated when Hunter said, "We see a real educational opportunity. . . . Alfred State College does not take a political stance. . . . You really don't need the support of institutions. You don't need to be afraid." He concluded, "We at Alfred State College are here to help with questions in the appropriate fields of technology and appropriate areas of education." Leaning over to Betsy, Steve muttered, "Who the hell does he think he is to tell us not to be afraid?"

Hunter's tepid statement disappointed Ostrower. He had hoped that the educational institutions would oppose the dump out of self-interest, if for no other reason.

One of the legislators asked about Alfred University's position. John Hasper read a letter from president Edward Coll, who, like Hunter, emphasized that the university could not take an institutional stance. Instead, Coll offered technical "expertise to serve the needs of interested parties in this debate. We will be pleased to provide professional guidance on both the merits and the hazards of the Allegany County sites and will make this expertise available to you and others. Alfred University is one of the national leaders in research on ways to create safe and secure methods of disposing of radioactive wastes and several of our faculty members are conducting research on this problem."

Hasper turned to provost Rick Ott, who represented the university at this meeting, and thanked him for the offer of help. "I understand that it wouldn't be appropriate for the university to take an institutional position. I'm really interested in getting the individual expertise from your faculty."

Betsy Myers seethed on the edge of her chair and barely refrained from shouting, "We're working hard to keep this dump out of the county and you're sitting up there pretending to be experts on safe ways to store nuclear waste." She grumbled to Steve, "They're so pompous and arrogant, pretending to be superior to us. We've been collecting information and studying the issues, yet they think they know everything. It galls me that Hasper's thanking them for coming to a meeting!"

Steve was equally incensed. The siting commission, he knew, would use the institutions' neutrality to crack open the county. They would hold seminars on the best way to build a nuclear dump and probably give the School of Engineering large amounts of money to design it. He found Coll's emphasis on the university's nuclear expertise chilling and saw how easily they might become complicit in siting a nuclear dump, while maintaining a guise of scientific objectivity. Both he and Betsy were now convinced that the colleges and politicians would sell out the county.

After the meeting Steve Myers and the activists met in one corner and Gary Ostrower and the scientists met in another. To Steve Myers, Gary's technical committee was another form of garbage. He and the other activists vowed not to let politicians, scientists, and college administrators sell out the people. Steve said, "I don't know how we're going to stop it, but we have to keep the colleges in this county from becoming part of the siting process!"

Ostrower told the scientists that he would set up a meeting in about a week. Several others volunteered to be on Hasper's committee. Fearing that the committee might become unwieldy, he mostly took names and told people he would get back to them. One of the most persistent, however, was Irma Howard, a well-dressed, prim woman in her forties. "I'd like to be on your task force."

"What do you do?"

"I'm a biochemist at Houghton College," she answered.

"We already have a biologist on the technical committee and I don't want to have so many people that we can't operate efficiently," Gary responded.

"This is different. This is chemistry looking at biology and you *need* a biochemist." Irma surprised herself with her own insistence. With a very busy schedule, she had even been reluctant to come to today's meeting.

Like many others in the county, she hadn't initially thought that a "low level" nuclear dump was very serious. She had worked with radioactive materials in the past and assumed that they were relatively harmless if handled sensibly. Her husband, who taught history at Houghton College, however, smelled a rat and urged her to investigate. The Howards subscribed to Houghton College's ideal of Christian service in the world. They believed education should prepare people to become active in their communities. When Irma started studying the technical material about "low level" nuclear waste, she was astonished to learn that it contained extremely dangerous irradiated metals from nuclear power plants along with relatively harmless medical and research-oriented materials. Impelled by a strong moral sense, she decided to use her education and scientific skills to ferret out the truth.

"All right," Gary responded. "I appreciate your willingness to be on the committee. I'll let you know when and where we'll have our first meeting."

JANUARY 7, 1989 —

The day after the meeting in the county courthouse, Ostrower called Betsy Myers on the phone. "Hello, Betsy. Yesterday Steve said you're planning a public meeting next Friday at the Almond Fire Hall. John Hasper would like to speak."

"What does he want to say?"

"I don't know for sure, but he wants to talk about political strategies for fighting the dump."

"We weren't going to have lots of speakers. We've invited the West Valley Coalition to tell us about their experience with nuclear waste. We want to allow plenty of time for questions."

"John doesn't want to talk long. He just wants to say a few words to let people know that he'll use all his political muscle to fight the dump."

Betsy suppressed her inclination to tell Gary what she thought of the politicians and college administrators at yesterday's meeting in the county courthouse. "We're getting together tomorrow to plan the meeting. We'll let you know."

"Would it be okay if I come?" Gary inquired. "We can talk about coordinating our activities."

"No, I think it's best for us to meet by ourselves. But I'll let you know what we decide." She said goodbye and hung up the phone.

JANUARY 8, 1989—

Emerging leaders in the eastern and southern parts of the county met in the Myerses' kitchen on Sunday to plan Friday's public meeting in the Fire Hall. Driving to the meeting, Rich Kelley told his brother-in-law, "I hope this is important. I can't believe they'd call a meeting during the football playoffs! Let's duck out early. I at least want to see the second half of the Bills' game."

As people were gathering in the Myerses' kitchen, a thin, wiry man in his late twenties, wearing a leather medicine pouch around his neck and sporting a long pony tail and scraggly beard, showed up at the door. He handed Betsy a bag of doughnuts and introduced himself. "Hi. How're you doing? I'm Jim Lucey and I've been reading about the nuke dump in the papers and saw that you're organizing against it."

Startled, Betsy invited him in, got out a plate for the doughnuts, and put them on the kitchen table. As people helped themselves, Steve asked Lucey, "Where're you from?"

"My family and I bought an old farmhouse next to state forested land near Belmont. We're going to build a house there. I've been doing some logging and construction work."

After a few pleasantries, Steve suggested, "Why don't you give me your phone number and I'll let you know what we decide. We're just starting to talk about it now."

"I'd like to stay," Jim said.

"But we don't know who you are," countered Steve, who was concerned that the fledgling movement convey an aura of respectability. "This is a private meeting for people who've organized groups in the area. We'll have to ask you to leave."

Jim was amused at this role reversal. He was used to being in protest movements where people looking like him were suspicious of "straight," middle-class folks. He noted that most of these people weren't even wearing jeans. "I fought against the nuclear power plant at Seabrook with the Clamshell Alliance. I was maced there, trying to cut through the fence. You'd better believe I'm committed to fighting the nuclear industry!"

Jim's words did not, however, allay Steve's fears about tarnishing the dump protesters with a "radical" image, especially in a rural county that votes overwhelmingly Republican. Another vague fear nagged at the back of Steve's mind. He had read about governmental infiltration of protest groups and how the nuclear industry had played hardball with people such as Karen Silkwood, who had blown the whistle on safety concerns at a nuclear power plant. Shortly thereafter she was somehow exposed to a lethal dose of radiation.

While Steve searched for words, Jim smiled and said, "Listen, you took my doughnuts. I'm staying!" Sandy Berry, recognizing an ex-hippie comrade, laughed and told Steve that he should let Jim stay. Betsy looked at Steve, shrugged her shoulders, and smiled.

Steve acquiesced, "Okay, I guess you're in."

Everyone expected Steve to continue, so he talked about what was on his mind—Hasper's meeting in the courthouse. "If those people start speaking for Allegany County, we've lost the fight. They won't do anything but talk."

"I'm still angry at their arrogance," interjected Betsy. "Those university know-it-alls and political big shots sat up front and were so condescending. They acted like they had all the answers. Imagine Gary Ostrower telling me that Steve couldn't come to the meeting! We've been doing more to organize people in this area than anyone." Betsy smiled, "Steve crashed the meeting anyway as my escort."

"Does Hasper really oppose the dump?" wondered Rich Kelley. "I know he's saying all the right things, but he could be in the pay of the nuclear industry."

"You might be right," Jim Lucey said. "Republicans have been pretty cozy with the nuclear industry."

Betsy scowled. "Can you believe that Ostrower called me yesterday and had the nerve to ask about coming to this meeting?"

"I can't believe Betsy was so polite to him on the phone," Steve added.

Betsy walked over to the counter with the empty doughnut plate. "We need to stay in contact with him so we know what they're doing. Gary asked if Hasper could speak at the Fire Hall. Do you think we should let him?"

When no one answered, Rich Kelley asked, "What do you think about the technical committee that Hasper's putting together?"

"They're fools," answered Steve. "That's exactly the game the siting commission wants us to play. They want to keep us talking about all the technical stuff. It's clear that's Alfred University's game too. Wasn't it cute how Coll offered the faculty's nuclear expertise? Neutrality, bullshit!"

No one said anything and Steve fumed, walking over to get a glass of water. "They'll try to get money for their so-called expertise and the siting commission will be more than happy to use the university to shove the dump down our throats. And who's going to be on Hasper's technical committee? So-called objective scientists from the university!"

"You might be interested in knowing," said Glenn Zweygardt, a sculptor at Alfred University who was also a leader of the protest group in Alfred, "that Tom Peterson has circulated a petition among faculty, asking the administration to reverse its neutrality and oppose the dump. More than half of the faculty have already signed it. He's going to send it to each of the trustees and to the newspapers."

"Who's Tom Peterson?" Steve asked.

"He teaches comparative religions at the university and is interested in Native Americans. He's very committed to stopping the dump."

"Is there any chance he'll succeed?"

"Your guess is as good as mine."

"We should definitely have him read his letter or give a short statement at Friday's meeting," Steve said. "Nothing's more important right now than neutralizing the university."

"I know him," Glenn said. "I'll ask him to come."

"Should we let Hasper speak?" Betsy asked again.

"Why should we?" questioned Rich Kelley. "He's already had his meeting."

"That was my first reaction," Betsy said. "But maybe we need to smoke him out. We can't let him become part of the university's so-called neutrality game. We've got to keep him responsive to the people."

"That's right," interrupted Glenn. "Let's be smart like a fox. Let him speak. He can't say anything that would hurt us, and it would be good to get him on record again."

They talked for several hours more, putting together an agenda for the meeting. Jim Lucey volunteered to contact the printer about publicity fliers. Rich Kelley missed the football game, but the Bills lost anyway.

About 6 P.M. while the group was writing press releases and designing fliers, the phone rang and Betsy answered it. She turned to Sandy. "It's the kids. They're hungry and want to know what to eat." The two nine-year-olds had been hanging out at Sandy's house all afternoon.

"Tell them to open a can of soup," suggested Sandy.

Five minutes later the phone rang again. Betsy once more turned to Sandy. "The kids can only find a can of mushroom soup."

"That should be okay. Tell them to mix it with a can of milk and heat it up."

A few minutes later the phone rang again. One of the kids said, "We can't find any matches to light the burner."

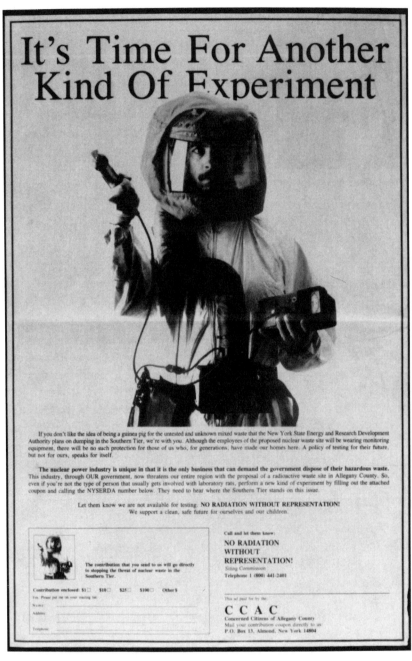

Ad Campaign © Steve Myers 1990; all rights reserved. Reproduced courtesy of Steve Myers Studio produced for CCAC's advertising campaign.

Suddenly Betsy became alarmed. "Don't do anything more. We'll be right there and get you something to eat." Hanging up the phone, she turned again to Sandy. "Whoa! Time out! The kids can't find matches! What on earth are we doing? We're so crazed, we're not even paying any attention to the kids. This is unbelievable. We haven't even taken time to eat anything but doughnuts. Let's break this up and go get some pizza."

The children of the activists would be eating a lot of canned soup and pizza during the next few weeks.

JANUARY 13, 1989—

The meeting at the Almond Fire Hall drew a crowd of seven hundred people, mostly from the eastern part of the county. Out in the hallway people set up tables to hawk "Dumpbuster" T-shirts and "Bump the Dump" bumper stickers. The Almond citizen's group sold copies of *Forevermore*, a series of articles in the form of a Sunday magazine supplement from the *Philadelphia Enquirer*, exposing a history of unconscionable negligence in storing nuclear waste. A pronuclear group, The New York State Low-Level Waste Generators (NYSSLWG—pronounced "Nissle Wig"), came down from Rochester. Sandy Berry had mistakenly thought she was talking to an antinuclear lobby and invited them to the meeting. The activists, though chagrined about discussing their plans with a pronuclear lobby, nevertheless allowed the group to set up a table and pass out its literature.

The story told by speakers from the West Valley Coalition galvanized the crowd. The Coalition's members had consistently kept the gross mismanagement of radioactive waste at West Valley in the public consciousness. Arguably, the tale of horror in neighboring Cattaraugus County had already predisposed the people of Allegany County to be suspicious of the rosy picture being painted by the siting commission.

The small village of West Valley lies about forty miles west of Allegany County's border. Near the village is an abandoned plant for recovering uranium and plutonium in "spent" nuclear fuel for reuse in nuclear power plants. This reprocessing plant was owned by New York State but had been operated between 1966 and 1972 by a private corporation. After the operation proved to be an economic failure and ecological disaster, the private company walked away, leaving the state of New York with an environmental mess of monumental proportions.

The state had neither the expertise nor the money to handle the witches brew of nuclear waste that was concentrated in an aging 600,000 gallon tank, nor did any technique exist for solidifying or encasing the waste. To make matters worse there were two burial grounds on the site for "low level" nuclear waste that presented serious environmental difficulties. Nuclear material had leaked and migrated from at least one of the clay-capped disposal trenches. In 1975 a trench had filled with water and overflowed, either due to a faulty clay cap, spring water, or both. An even more serious possibility existed that part of the burial grounds might wash into a series of creeks that eventually flow into Lake Erie.

In 1980, urged by a group of citizens and political leaders, the U.S. Congress agreed to take charge of the project in order to "demonstrate" the proper way to clean up nuclear contamination. They would fund 90 percent of the project, with the state of New York picking up the remaining 10 percent. By the end of 1998 they had spent more than one billion dollars and were nowhere near finishing the task. Some estimates were that it would take another eight billion dollars to clean up the site.

After the siting commission announced the potential dump sites, members of the West Valley Coalition began sharing their expertise with local people throughout Allegany County. Here at the Almond Fire Hall, Carol Mongerson talked about the false promises that were made by the state and nuclear authorities to gain the cooperation of local citizens. "They promised us 'It won't leak' and, 'It will make you rich.' Well, we're not rich and it did leak; there's plutonium in a dam three miles away."

She explained that "low level" waste included some very highly radioactive materials, some of which would be around for hundreds, and even thousands, of years. She emphasized the serious responsibilities entailed in accepting a nuclear waste dump in a community. The West Valley Coalition had concluded, she said, "that the only thing to do with it is to keep it retrievable. We'll have to watch it for a long time—we'll have to teach our children to watch it—but we really do not have a choice."

Most importantly, she warned, the siting commission would try to get people squabbling among themselves. She predicted that people would start criticizing those who were more active as rash and those less active as timid. "If there are disagreements among you, ignore them and fight the nuclear dump in your own way, without criticizing those who follow other paths. You will not win if you start fighting among yourselves."

Steve Myers was grateful that Mongerson had so clearly stated the dangers of nuclear waste. Maybe now people in the county would see that he was no unreasonable alarmist.

Assemblyman John Hasper picked up on Mongerson's theme. "We have to stay cool, stay smart and stay committed so that the commission will not be able to separate our points of view. Benjamin Franklin said it best at the signing of the Declaration of Independence and it holds true in this situation. 'Gentlemen, if we don't hang together, we most assuredly will hang separately.' "

Steve Myers had worried all week that apathy would keep many people from attending. Now, with seven hundred people tightly packed into the large room, he could relax. The meeting couldn't be going better. Hasper had again made his position against the dump absolutely clear and Steve would publicly hold him accountable.

After Hasper spoke, Steve introduced me. "Tom Peterson teaches world religions at the Alfred University and is putting pressure on the administration there to take a stand against the nuclear dump." As I came forward, Steve recalled a telephone conversation we'd had that morning.

"Glenn tells me that you don't want to speak at Almond," Steve had said.

"I really don't have anything to say. I'm simply trying to get the university to change its position," I replied.

Steve wondered whether he could be completely candid and explain why he wanted me to speak out. "Look, nothing's more important right now than to neutralize the university. The siting commission will try to suck the university into cooperating with them."

"You're right. But what does my speaking in Almond have to do with all this?"

Steve Myers decided he'd have to be direct. "You'll make the community aware of the university's position and that may embarrass them."

Had he gone too far? Steve wondered. Would the professor back away from such a direct confrontation? Steve held his breath. No one said anything for a couple seconds.

"Why not," I said, "if you really think it'll put pressure on them?"

That night at the Almond Fire Hall I concluded my statement, "The faculty are very upset that the university has not taken the lead in opposing the dump." Steve rose to his feet and led a surprisingly loud applause. Clearly I had underestimated how strongly local people felt betrayed by the university's neutrality.

During the week after the meeting in the Almond Fire Hall, Alfred University did in fact change its position. The provost and president decided to back the large majority of faculty who wanted the school to oppose the dump. They told newspaper reporters that their initial statement had only been "provisional." Provost Ott also put forth an intellectual rationale: When there were already so many contaminated nuclear sites in New York State, it would be wrong to create more.

Houghton College issued its own statement opposing the dump a few days later.

❖

Over the next three months, hundreds of meetings were held all over the county. Representatives of the many citizens' groups met in Angelica to form a county-wide organization, becoming the "Concerned Citizens of Allegany County" (CCAC). Steve Myers became president and Rich Kelley vice president. In addition to secretary and treasurer, all sorts of other offices such as action coordinator, outreach coordinator, and publicity coordinator were invented so that other county leaders had official positions in the newly formed organization. Each of the ten local groups would have autonomy, yet pool their resources and coordinate their activities.

Ostrower assembled the technical committee at provost Ott's house in Wellsville, the county's only industrial center with a population of slightly over five thousand people. Holding a meeting at the provost's home inflamed Steve Myers's suspicions that the university was still playing a double game, even though it had just gone on record opposing the dump. When word leaked back to Steve that Gary Ostrower had again emphasized that the scientists should approach the siting commission's report with complete and total objectivity, he stewed all the more.

Steve and Gary watched each other carefully over the next three months. Gary respected Steve's energy and organizational skills, but worried that his intemperate rhetoric would give political leaders in the county an excuse for doing nothing. Steve admired Gary's vitality, but fretted that the technical committee would inadvertently sell out the county. When Steve discovered that the next meeting would be at Gary's house, he crashed it. "After all," he later explained, "the technical committee was planning to speak for the whole county and the citizens had a right to know what was going on." He shared an even darker thought with Betsy. "They will know they're being watched." Although irritated at his presence, Gary bit his tongue and said nothing.

Gary's annoyance eventually gave way to grudging respect for Steve's political savvy and dogged determination. When honestly thinking about it, Gary even had to admit that he was, in fact, playing a dangerous game that could backfire if the siting commission opened up serious dialogue with scientists in the county.

Steve's mistrust was finally overcome on March 31, 1989, at a meeting in Fillmore, a rustic farming community in the northwestern corner of the county that had hardly changed since the nineteenth century. Gene Hennard, owner of the local feed mill and president of the northernmost CCAC chapter, had, against Steve's strong admonitions, invited members of the siting commission to Fillmore. The other county leaders had already developed a unified strategy to have no more dealings with the commission, because they did not want to legitimize the process or signal any willingness to make a deal.

Never fully integrated into the county-wide movement, Hennard rejected this strategy, feeling it was more important to challenge the siting commission directly. No CCAC chapter wrote more letters to politicians or called the siting commission more often than did his group. Gene personally called them a couple of times each day to pester them with questions.

When Steve called Hennard to ask him what he was doing, Gene asserted his independence. "Let's face it," he said, "it's a crowd getter. If you have top performers, you're going to get a good crowd."

Sputtering, Myers tried to explain why no one should invite the siting commission into the county. It was dangerous, he said, to open up lines of communication with them. This argument made no more sense to Hennard than it did to Gary Ostrower whose technical committee was initiating a scientific dialogue with the commission. Gene felt the meeting would allow his people to blow off some steam and hassle the state even more. "Anyway, it's a done deal. Kathleen McMullen and Richard Wood have already agreed to come."

Wood was one of five members on the siting commission. He had worked for Niagara Mohawk at their nuclear power plant in Oswego for more than twelve years and now conducted research, funded by the nuclear industry, at Syracuse University. Wood feebly argued that since less than 10 percent of his salary came from his old employer, and since it was indirect, there was no conflict of interest to his serving on the commission. McMullen, a physicist, was the commission's liaison with the county.

Although Hennard's meeting was not publicized much outside the northern parts of the county, leaders in the protest movement all showed up anyway. Gary Ostrower decided to give the recently completed technical report to Wood during the meeting. When he got up to present the sixty-eight page report, many of the CCAC leaders were uneasy, and a few heckled him. One woman blurted out, "Who gave you and Hasper the right to speak for the county!?"

Ostrower glanced at her and walked confidently up on the stage. Gradually, as he read excerpts from the report's introduction, the tension eased. "For all the good intentions of the siting commission's plan . . . political considerations have already corrupted a technical solution to the waste problem." The report chastised the commission for its "bewildering refusal to select a disposal method in advance of selecting a site," and questioned "the meager funding that makes failure virtually inevitable."

The mood of the audience changed to exuberance when Gary read the technical committee's conclusions. "Some of what you will read suggests that New York State's plan for low level waste disposal needs to be rethought, some suggests that the application of siting commission criteria . . . is so seriously flawed as to constitute a public injustice, and some suggests" that sites in Allegany County "simply do not meet the requirements . . . as established by your commission." Myers breathed a huge sigh of relief. After the meeting he began mending fences with Gary by personally thanking him for his work.

A scathing scientific critique of the statistical methodology used by the siting commission in choosing the sites formed the heart of the report. What attracted the media, however, was Irma Howard's imaginative, yet quirky, argument that no nuclear dump should be built where cluster flies existed. Cluster flies, unlike ordinary house flies, reproduce, not in sewage or garbage, but in earthworms. Rampant in Allegany County, these flies have an uncanny ability to violate human structures. No matter how clean a nuclear waste facility might be, these organisms would be potential vectors of its radioisotopes. Henceforth Dr. Howard, a nationally well-known biochemist, became affectionately known in the county as "the cluster fly lady."

The county got no response for eleven months, long after the commission had narrowed down scores of potential areas in the ten counties to five specific sites, three of which were in Allegany County. Although the report's scientific conclusions never got a hearing at the siting commission, it became a powerful political document that scientifically confirmed the suspicions of people in the county that they had been targeted for political rather than technical reasons. Allegany County was a rural, Republican, economically depressed enclave in a heavily urbanized, wealthy state that had a Democratic governor and Democratic Assembly. As newspaper reporters summarized the technical committee's geological, biological, and statistical conclusions, the report served to foster people's certainty that the proposed dump was one more instance of a quick fix for the nuclear industry—a twenty-four million dollar solution to a multibillion dollar problem.

❖

JANUARY 26, 1989—

The meeting that most significantly changed Allegany County's perception of itself took place two months earlier in a high school gymnasium. Crowds had grown as the siting commission held public meetings in the targeted counties across the state, moving from east to west. Only a couple of days earlier, three thousand people had confronted the siting commission in Chenango County in central New York. Steve Myers worried that far fewer people would show up in Allegany County. He and other activists redoubled their activities to get people to attend.

In his regular newspaper column John Hasper urged "all those concerned to come out and . . . tell the governor's commission how we feel." Leaders in CCAC contacted newspapers and radio stations. Posters were plastered in every grocery store, restaurant, and bar in the county. Gary Ostrower said on the radio, "Don't come to the meeting if you want the dump." An editorial in the *Wellsville Daily Reporter* proclaimed, "Allegany County needs you in Belfast. Someday, people will say, 'Remember the Belfast nuke meeting?'"

A cold rain was letting up at 6 P.M. as a steady stream of cars crawled toward the tiny village of Belfast. The slick pavement glistened under a mile-long glare of headlights. A January thaw had turned snowy fields into mud and dirt roads into squishy ruts of clay. With the thermometer hovering just above freezing, people feared roads would become sheets of ice before the night was over.

At six o'clock the sheriff posted deputies at the school door to prevent more people from entering. A northerly wind whipped water into people's faces and up their sleeves. By the time the meeting started an hour later the rain had stopped, though a bone-numbing fog saturated the couple of thousand people gathered outside.

Not only was the gymnasium packed, but hallways and other rooms were jammed. Hundreds viewed the proceedings on TV monitors in the cafeteria. Loudspeakers blared rhetoric into the school yard. Images of a warmer world flickered on a TV monitor in a bus garage where shivering people huddled around a huge heater, vainly challenging the chilly dampness blowing in through open bay doors.

Ten simulated gravestones with the names of the sited counties lined the walkway to the building. A Buddhist nun from Japan, witnessing against the horrors of the nuclear age, spent the entire six hours slowly beating a drum, one beat every couple of seconds, while her companion made and passed out origami cranes, symbolizing peace. A Native American, probably a Seneca from a reservation seventy miles away, handed out leaflets, arguing that humans were caretakers of the earth, not owners of the land. Just as people were getting off a bus from Alfred University, an electrician who worked there stood up and told the students, "I'll never forget you joining us tonight. You're here in the county for only four years. You'll be gone when they put the dump here. I want to thank you for your effort."

The sheriff's department estimated that five thousand people had come to Belfast that night, nearly 10 percent of the fifty-two thousand people who lived in the county. Nothing was said that night that would change any minds, but commitments were strengthened. For the first time in the history of Allegany County people from all walks of life showed up to protect their families and land—village dwellers and farmers, intellectuals and high school dropouts, business people and professionals, workers and those on welfare, Republicans and Democrats. Even the Amish, who normally avoid political meetings, had come. John Hasper had invited them and saved them seats near the front. When the fourteen elders walked into the meeting moments before it began, they were met by thunderous applause. Now the county was complete. Throughout the evening they said nothing; mere presence was their statement.

Banners proclaiming "No Dump Here" and "Mourn the Death of Allegany County" mingled with the high school's more permanent slogans, "Our J.V. Will Stomp You" and "Our Varsity Is Red Hot," on the gym walls. A woman from one of the CCAC chapters hawked glow sticks for a dollar apiece, while another group peddled Dumpbuster sweatshirts for twenty and T-shirts for ten. People conversing with their neighbors raised their voices to compete with the sound echoing off concrete walls and wooden floors. Damp odors steamed off wet clothing and fidgety people chanted "No Dump" while waiting for the meeting to begin.

The siting commission presented their charts, graphs, and drawings of concrete bunkers, filled with nuclear waste, to the restless crowd. Slogans from hecklers punctuated the wooden, lackluster presentation. "We ain't tak'n it!" "Go back to Albany!" "We're not expendable!" After a particularly tedious monologue on options for storing nuclear waste, a woman shouted, "When do we vote?" The siting commission was able to say its piece only because Steve Myers, impeccably groomed and wearing a dark suit, flashed a hand-printed sign, reading "Quiet Please."

For another five hours following the commission's presentation, people expressed their profound revulsion against nuclear trash and their commitment to fighting the state. John Hasper, the second to speak, told the commissioners that they were facing "the initial efforts of the combined political, economic and social forces of the fifty-two thousand people in Allegany County." Many people will explain, he declared, "why this dump is inappropriate and unacceptable."

When people are faced with emotional issues, they seek to find metaphors to express their inchoate, but heartfelt rage. One woman had created a huge hand-held puppet called "The Mutant," a cyclops with a vertical mouth, pock marks, and an open head with a brain pouring out of it. Painted in grotesque Day-Glo colors and wearing a gaudy turquoise skirt that hung down to the floor, the puppet bore a sign asking, "Will this be our destiny?" One man wore a grim reaper costume, while his wife chanted, "Death comes! Death comes!" during pauses in the speechifying.

The most successful metaphors of the evening were those that alluded to the American Revolution. After the commissioners' presentation Steve Myers was the first to speak. For days he had agonized about what he should say and asked others for suggestions. During one of their many exploratory conversations, Gary Ostrower had suggested that he look at the Declaration of Independence for inspiration. Now Steve spoke about the "tyranny of the majority," who would create an "atomic wasteland the size of a small city." Then, without introduction, he began reading the Declaration of Independence. Thunderous applause greeted the phrases, "Consent of the governed," "life, liberty and pursuit of happiness," and especially "the right of the people to alter or abolish" government itself. As he read from this ancient document, Steve swallowed his own emotions with difficulty, intensifying the power of the words.

The revolutionary theme reverberated throughout the evening in speech and symbol. A woman who wrote children's stories wore a tricorn hat. A professor brought a revolutionary "Don't Tread on Me" flag, with a coiled rattlesnake on it. A mother with three children asked, "Why shouldn't people decide the future of nuclear power democratically?" Amo Houghton, the United States Congressman who represented the county, asked rhetorically, "If we don't make it, and if we don't use it, and if we don't want it, why should we have it?"

Most of the people who approached the open microphones were not used to speaking in public, but felt impelled to express their defiance and to bear witness to the love of their families, their land, and their rural lifestyle. Volunteer firefighters questioned the emergency planning. If a disaster happened, one said, "we're going to take one look at it and run like hell." A construction worker wondered if the siting commission knew the half-life of cement. Gene Hennard, the owner of a feed mill in Fillmore and irascible leader of the northernmost chapter of CCAC, wondered why nuclear power plants were allowed to operate before developing a disposal system. After all, he argued, we can't put toilets in our houses before we build the septic system.

A farmer, whose family had lived on the land for seven generations, explained that water flowed in all the cardinal directions from streams near his land, some of it ending up in Lake Ontario to the north, some of it in the Mississippi River, and the rest in Chesapeake Bay. As he traced the flow of water past trees and through meadows, people felt his intimacy with the land. Concluding, he told the commissioners that "God so much loved this country and he so much wanted us to share it with the rest of the eastern United States, that he did it by means of water." A momentary silence greeted his final rhetorical question. "Do you people really want to defy God?" Then foot-stomping, whistling, and applause.

While people were testifying about their love for the land, the siting commission sat unmoved, uncaring, and bored, like atheists at a revival meeting. Testimony that seemed crucial to people in the county was to them irrelevant. An elderly engineer, hands folded over a large gut, slept through much of the meeting. At times the commissioners tried to suggest that people were hysterical and unjustifiably frightened because they didn't know enough about radionuclides nor understand that the "storage facility" would be "state of the art." At such times anger nearly overwhelmed the general civility of the meeting.

When John Randall, head of the state agency authorized to run the dump, walked through the crowd on his way back from the bathroom, he was accosted by Stuart Campbell, "How can you look at yourself in the mirror?"

Randall turned and looked disdainfully at the man sporting a scraggly beard and wearing an old pair of jeans. "The best thing you could do is educate your children," he advised, unaware that Stuart was a professor of history at Alfred University with a son who was studying architecture at Rice University. Campbell would later become one of the founders of the nonviolent resistance movement in the county.

Near the end of the meeting someone finally asked the people on the raised platform to give biographical sketches of themselves. Only two of five commissioners were present. Richard Wood said that he had worked for the Niagara Mohawk nuclear plant for more than twelve years. He failed to mention, however, that he was still indirectly receiving money from his old employer to do research at Syracuse University. David Maillie talked about his long-time involvement in research in the Department of Nuclear Medicine at the University of Rochester Medical Center. He failed to mention that the State Department of Health had cited his department for more than forty violations in the last two years, including illegal dumping of radioactive materials in the Genesee River, the loss of radioactive material, and contaminated desk drawers and trash cans. When a reporter for a Syracuse television station later asked him to justify the dumping, he said it was safe. "After a period of time materials in the Genesee River or a sewage system get diluted."

Gary Ostrower was one of the last to speak at the meeting. He told the commissioners that they should "have the guts to tell the governor . . . that your assignment is wrong." Tell the governor, he said, that the opposition is so strong and the plan so faulty that it should immediately be stopped in its tracks and sent back to the drawing board. Richard Wood could not resist the bait and told Ostrower that he felt the plan was a good one. "I think the job is an appropriate job and we're going to do it!"

At no time in the meeting were catcalls so loud. One tall young man stood up amidst all the cries of defiance and got everyone's attention. Shaking his fist at Wood, he shouted, "It's hotdogs like you that would make me stand right at the county line and say, 'Hey boys you ain't coming in here at all, ever!'" He would later be a monitor at a roadblock in West Almond, preventing the technical team from entering the site.

That night I also spoke, modeling my words on those of Winston Churchill. I told the commissioners, "We will fight this decision every step of the way. We will fight you in the halls of science. We will fight you in the law courts. We will fight you in the legislature. And if necessary we will fight you in the streets." The crowd affirmed their resistance with thunderous applause. I made a decision then to involve myself in nonviolent resistance, and I would help found the Allegany County Nonviolent Action Group a few weeks later.

2 Birth of Nonviolent Resistance

Civil Disobedience. Thoreau wrote about it, Gandhi practiced it. . . . It's an American right, and no single individual, no government agency, no big business concern, and no political extremist group can take it away from us. . . . It's easy to give a buck for a pin, sign a petition that is only half read, or donate a few dollars to an organization that has dedicated its proceeds to the fight. It takes a lot to link arms with your fellow citizens and wait to be arrested.

—Kathryn Ross, of the *Wellsville Daily Reporter* in column "My Home Town" on June 5, 1989

MARCH 19, 1989—

GARY LLOYD SAT ALONE in a corner of his basement. The last of twenty men, most wearing camouflage, had just left and it was nearly midnight. They had formed a clandestine group called the Allegany Hilltop Patrol to fight the nuclear dump. Quickly agreeing not to do anything to harm people, they had begun talking about ways to sabotage drilling rigs and bulldozers. People in other movements might call it "ecotage," but the local men meeting in Lloyd's basement simply referred to it as "vandalism," albeit vandalism for a good cause.

Although Gary had known most of these men for a long time, he was uneasy with the group's size. Almost everyone he had called had brought two or three friends. All evening Gary had listened and watched, as though he were observing animals in nature. An avid bow hunter, he was used to interpreting the moods of deer by observing their tiniest movements. These guys were, he concluded, deadly serious. A few, however, seemed a bit too eager to blow something up, while others appeared overly nervous.

The mere thought of a nuclear dump in Allegany County was a personal assault on Gary's world. As a little boy spending summers and weekends on his grandmother's farm, he had freely roamed through the wilderness and played in the creeks. When he was five years old, he learned to fashion spears and spent hours trying to catch sculpins, tiny minnow-like fish that scooted along creek bottoms. His spear raised over the water, he would wait motionless at the bank for just the right moment to strike. Eventually he learned their habits and could predict their movements. For Gary that was hunting—focusing on an animal until he knew its habits so intimately that he could take its life.

As an adult he hunted deer the same way. During the winter and summer months he would hide in the woods, camouflaged, watching the movements of bucks and does. Before bow season began in the fall, he would pick out the cagiest buck, who interested him because he was so unpredictable. He wouldn't shoot any other deer that season. Some years he would get his buck; often it outwitted him. Hunting for Gary was a time to harmonize with nature and think like his totemic animal, the deer. Poisoning this natural world with nuclear trash was sacrilege. Gary felt morally justified to explore all options to keep the dump out of the county—including vandalism.

Now, however, he began to have serious doubts about sabotage. How could they be so sure no one would get hurt? How would the people in the county respond to such acts of vandalism? Would vandalism divide people in the county, giving them an excuse to remain uninvolved?

Gradually, an idea formed, and Gary smiled to himself. Why not civil disobedience? Maybe the siting commission could be kept off the land without resorting to clandestine operations. The forty-six-year-old teacher remembered reading about nonviolent resistance during the Civil Rights Movement when he was in college. Just after he started teaching high school biology at Alfred-Almond Central School, friends who lived in Rochester and Buffalo used civil disobedience to protest the war in Vietnam. As an untenured new teacher in a rural, conservative county, Gary regretfully felt he could not join these protests. Why not try civil disobedience now?

The next day after breakfast Gary headed to Steve Myers's house. He admired the way CCAC had unified the county against the dump, even though he felt that legal battles and legislative lobbying would not save the county. If anyone would be able to get people actively involved in civil disobedience, Steve could do it.

"Steve, why not use civil disobedience against the siting commission?" Gary blurted out, when Steve opened the door. "We've got to become more militant. We should confront the siting commission when it gets here."

"I don't know," answered Steve, ushering Gary into the kitchen. "I'm not sure that we can do much until they bring in equipment, and I'm not sure people are willing to throw themselves in front of bulldozers."

"The siting commission will be coming to do windshield surveys in the targeted areas," Gary reminded him, sitting down at the kitchen table and accepting a cup of coffee. "There might be a chance to stop them." The siting commission had announced a week earlier that they would meet county officials and then drive around the sited areas in cars.

"What have you got in mind?" asked Steve.

"I'm not really sure. I don't know the specifics of their plans. Whatever we do, we need to be well organized—we can't just send people out to demonstrate. We've got to think through our options, and train people to act nonviolently."

Gary stroked his beard. "Steve, you could convince people to do this sort of thing."

"I don't know, Gary." On the one hand Steve was positive about civil disobedience, if it were well organized; on the other hand CCAC was already overwhelmed. He continued, leaving Gary to fill in the gaps of his thinking, "I've just put in a second phone line, because our phone's been ringing off the hook. We couldn't even make a phone call out. Now both phones are continually ringing. We're writing articles for newspapers, selling bumper stickers and T-shirts, calling political leaders in Albany, organizing rallies, and thousands of other things. I don't think we can take on something else."

"But this is really important." Gary thought of last night's meeting. "People are talking about guns and bombs, and neither of us wants that. We need to tap into that energy. These aren't the sorts of people who'll sit through long meetings talking about legal options, but they might get involved in civil disobedience."

"I've heard a lot of loose talk about guys picking up guns. But I doubt they're really serious."

"The ones who're talking probably aren't, but I know for sure that some people are forming underground groups right now, and they aren't talking."

"How do you know that?" Steve questioned.

"You don't want to know. Anyway I can't tell you. But it's really important for CCAC to create a mass movement that'll stop the siting commission nonviolently."

"I'm not at all against civil disobedience," Steve reiterated, "though I'm not sure that people around here will do it. Stuart Campbell called me a couple of days ago and suggested the same thing. I told him to come to the next CCAC meeting on April 10. Do you know him?"

"Not really, though I had his son in biology class a few years ago. He teaches history at the university, doesn't he?"

"Yeah. Why don't you come to the meeting, too?" Steve suggested.

❖

APRIL 10, 1989 —

CCAC met under a fluorescent glare in a cinder block room in the County Courthouse. During much of the meeting, David Seeger, a young lawyer from Buffalo, explained the danger that the county faced from a temporary or "interim" storage facility for nuclear waste. The 1985 congressional law required that states take title to all the "low level" nuclear waste generated within their borders by January 1, 1993. It now seemed doubtful that New York would be able to build a permanent facility by the deadline, and the state might have to build a temporary storage facility. Plans were afoot to put it at one of the finalist sites.

"A temporary dump would not have to meet all the environmental requirements that have been established for a permanent one," Seeger explained. Recently hired by CCAC, he had previously worked with other environmental groups, including the West Valley Coalition. "Temporary dumps tend to become permanent ones. We have to get the governor to remove language in the certification statement allowing temporary storage of nuclear waste where they are considering a permanent facility."

"What's a certification statement?" someone asked.

"According to the federal law the state has to certify that it's making progress toward building the dump. I'm disturbed about some of the language that I've seen in an early draft. I've written a letter to the governor's office threatening a lawsuit if it isn't changed. 'Don't Waste New York' is also putting political pressure on the governor to change this. CCAC should join them." Don't Waste New York (DWNY) was a newly formed organization with representatives from all of the targeted counties.

As Gary Lloyd listened to Seeger, he became more certain that CCAC should adopt civil disobedience as part of its tactics. If the siting commission didn't even have to prepare an environmental impact statement before bringing in nuclear trash, then the legal battle might be even less important than he had thought. This seemed a perfect opportunity to present his case.

"I think we should consider some extralegal means for stopping this," Gary interjected. "We need to organize people to put their bodies on the line."

"We can't encourage mob violence," someone interrupted.

"Civil disobedience," Gary quickly replied, "is not something that's haphazard or helter-skelter. It has to be well planned and organized in a military-type way."

"What are you talking about—military!" a CCAC leader from the township of Ward exclaimed.

With his scraggly beard, rumpled hat, and camouflage clothing Gary looked the part of a "survivalist" in a paramilitary organization. "Are we all going to be walking around carrying guns in a couple of months?" Fleurette Pelletier wondered nervously to herself. An environmental activist who had long been fighting a proposed ash dump in the county, she had adopted the name "Concerned Citizens of Allegany County" for her group several months before the nuclear threat surfaced.

Gary wished he hadn't used the word *military*, but the damage was done. He could see that most of the people were upset. Stuart Campbell, a short, husky man in his late forties, briefly tried to salvage the situation. He explained that civil disobedience required discipline so that people could remain completely nonviolent. "Nuclear trash, like all other garbage, flows like water. It'll go where there's the least resistance. If people are willing to be arrested to oppose the dump, the siting commission will see we're serious."

"You're still talking about illegal activity," someone else nervously interjected.

"You've got to remember that these are tactics for a good cause," replied Campbell. "We don't have any economic or political power. Civil disobedience is one of the few effective ways that people can respond to injustice when there are no democratic remedies." He could see that most of the people in the room were threatened by the notion of doing any illegal activity so he decided not to argue the point further.

People said nothing for about a minute. Steve looked at Gary and shrugged. Dave Seeger cautiously stepped into the void. "There may come a time when civil disobedience is an appropriate tactic. But there are many other ways to resist the state. If they know we'll put up a strong legal battle, that can deter them. Since no place in western New York will be able to meet the requirements set up by the Nuclear Regulatory Commission for storing low level nuclear waste, I think we can win the legal battle."

Seeger paused, then continued, "Let me emphasize again that we must stop them from establishing a temporary dump that wouldn't have to meet environmental standards. If we can stop them from doing this, we'll be able to beat them in court."

Stuart Campbell stared at his hands. These well-meaning people, he thought, were being sucked into playing the siting commission's game. They were kidding themselves to think they could outmaneuver the nuclear lobby in Albany or beat the state in the courts. New York had battalions of lawyers and unlimited economic resources. The state wasn't going to be intimidated by one young lawyer from Buffalo, no matter how competent he might be.

"If CCAC gets involved in organizing people to break the law," Seeger continued, "your treasury could be threatened by injunctions and lawsuits. It could even derail the legal actions that we've taken so far." Looking at Steve, he continued, "You and the other officers of CCAC should not get involved in any illegal activities." He paused, and then added, "Legal and political battles aren't as glamorous as civil disobedience, but you need to fight those battles if you want to win."

Neither Gary nor Stuart wanted to create divisiveness within CCAC. While neither believed the county could ultimately win the court battle, Seeger's comments did make sense. CCAC, as currently constituted, had committed itself to legal and political activity. There was no point in trying to get reluctant people to organize civil disobedience, especially when they were so completely committed to fighting the state on other levels. Stuart remembered the words of Carol Mongerson from the West Valley Coalition: "Fight the nuclear dump in your own way, without criticizing those who follow other paths. You won't win if you start fighting among yourselves."

After the meeting broke up, folks drifted off to a bar located in the old Belmont Hotel a couple of blocks away. At the beginning of the twentieth century someone like David Seeger who had ventured into the county on business would have stayed there overnight. But with improved roads and faster automobiles Seeger could now have a beer with the leaders of CCAC, hop in his car and be asleep in his own bed in Buffalo by 1 A.M.

After World War II the hotel had become seedier, and its rooms were rented by the month to workers with temporary construction jobs or to elderly men without families. Dim lighting hid peeling paint and dilapidated furniture while spotlights above a beautifully carved bar highlighted its fine woodwork. After a few beers, young men and women could pretend that they were attractive studs and sexy broads in some fine establishment. Older men and women could forget their worries while complaining about politicians and bosses who were messing everything up. Only the realities of daylight, a hangover, and the need to get to work would finally break alcohol's magic spell.

Stuart and Gary found a table in a back corner of the bar and ordered a couple of beers. "I shouldn't have used the word *military*," Gary lamented. "It made them think of guns and paramilitary groups."

"That didn't make any difference," Stuart replied, rubbing his beard. "CCAC is committed to going the legal route. They could be right, but I really don't think we'll beat them in the courts. People around here can't afford the cost of a lengthy legal battle. That's why they targeted us in the first place."

Stuart didn't know Gary well. Now he wondered whether this guy dressed in camouflage could lead a nonviolent resistance movement.

Looking at the university professor, Gary wondered how far he could trust this guy, even though he was wearing faded jeans and a rumpled shirt. "How committed are you to civil disobedience?" he finally asked.

"I've never done it before, but I think it's the right thing to do here." Stuart paused and took a sip of beer. "Gary, I'm no activist. I'm a critical thinker. As a historian my perspective is quite radical, even cynical about who really controls us. This system is run by powerful economic interests like those of the nuclear establishment. They don't give a damn about the environment. They'll grind it up to make a few bucks. They'll grind up people too. Even though Karl Marx may have been wrong about a lot of things and his theories aren't fashionable right now, he was basically right about economic exploitation."

As he talked, Stuart realized how much he cared about this area of the country. As a historian he had been trained to stay emotionally uninvolved while exploring how economic systems, intellectual ideas, and people's struggles shaped societies. Now, however, he was confronted with those historical forces right in his own back yard. If he avoided this battle, he wondered whether he could ever look at himself in the mirror. Although he'd probably deny it, his radicalism was deeply rooted in idealism that involved protecting the environment and fighting for powerless people against corporations willing to destroy them for profit.

"You asked me how committed I am to civil disobedience," Stuart repeated. "I think it's the only thing that's got a chance. The nuclear industry owns the politicians, and the state owns the law courts. I'm afraid everyone will get mired down playing the siting commission's game. Civil disobedience has sometimes helped change the ground rules in favor of the little guys. If we can create a strong movement, they'll be forced to react to us. They want us to play their game. We've got to change the rules and even the game itself."

Gary took a sip of beer, waiting for Stuart to continue. Stuart looked into Gary's poker face and said, "I'm really talking as a historian. I'm not sure whether it'll work in this situation. I'm neither a leader nor an activist. That's why I'd hoped to get Steve Myers and CCAC interested."

Stuart looked into Gary's eyes and wondered whether this man would be able to create such a movement. He waited a few seconds for Gary to speak and then said, "I guess I'm willing to help out, if you decide to organize nonviolent resistance." Both men were comfortable with silence. Stuart swilled the beer in his glass, took a sip, and waited for Gary to respond.

"I don't have any personal experience with civil disobedience either," replied Gary. He cautiously added, "If it's going to happen I guess we'll have to do it. I don't know whether I should tell you this. I know some guys who're thinking about sabotaging the nukers' equipment. I've been thinking about joining them. I'm not against doing that sort of thing, but to tell you the truth I really don't have the stomach for it—at least if there's any other way."

"I know I'm not interested in doing that," Stuart said. He didn't, however, want Gary to think he was morally opposed to such action, and added,

"My wife Sally would probably be more apt to get involved in that sort of thing. I'm not against it in principle. I just don't think it'd accomplish much in this particular situation. The state knows all too well how to deal with violence. They'd paint a negative picture of the whole movement and frighten people into silence. From the reaction at tonight's CCAC meeting, I wonder whether civil disobedience would even work, but it might have a chance."

"Are you willing to work with me to organize civil disobedience?" Gary asked pointedly.

Stuart was never one to make a rash decision. "Let me think about it. I came tonight to make my pitch to CCAC. I'm not sure how involved I want to get. I'm no leader. That's just not my character."

Both men finished their beers in silence. Gary finally spoke. "We need to find a few people who know something about civil disobedience."

"I have a friend who teaches religious studies at the university who was arrested for blocking an army induction center during the Vietnam War," Stuart interjected.

"I know someone else," Gary mused, "who was arrested many times as a labor organizer during the 1950s. Why don't we get a few people together at my house next Wednesday night. Let's keep it small at first."

❖

APRIL 26, 1989 —

Gary Lloyd kept all of his hunting and fishing equipment in the large basement of his house, which he had turned into his special room, a male retreat. People coming in from the outside were immediately struck by a huge wallpaper mural of a buck and two does in the woods, which covered half the far wall. An old refrigerator, filled with soda pop and beer, stood by the door. A hot tub that Gary used when he came home from hunting occupied a far corner, and a pool table took up considerable space in the center of the room. Near the mural was a large, round table with plush leather chairs, crafted from deer hide. A casual observer would think the eight men who were sitting around the table were probably playing poker.

In ages they ranged from thirty-two to sixty-two. Four had been arrested in earlier protest movements. Dave Davis, the eldest, now a teacher at Alfred State College, had been a union organizer in the International Ladies Garment Workers' Union in the 1950s and had been arrested in at least four states—New Jersey, Missouri, Ohio, and Mississippi. Jim Lucey, who had earlier forced his way into the inner circle of CCAC, was the youngest. He had been arrested during protests at Seabrook, a nuclear power plant on the coast of New Hampshire. Mark Kurath-Fitzsimmons had participated in some Earth First! civil disobedience protests, and I had been arrested for blocking the entrances to a military induction center in Oakland, California, during the Vietnam War.

In addition to Gary and Stuart there were two others: Walt Franklin and Bill Castle, who were known locally for their strong commitment to living in harmony with the environment. Walt had just published the first of sixteen issues of *The Stone and Sling*, a photocopied magazine containing poetry and articles that celebrated nature and denounced the corporate greed that was wrecking the environment. Bill was an artist and engineer, who had built an energy-efficient home at "Pollywogg Holler" and ran a bed and breakfast there.

Gary began the meeting by reminding everyone that the siting commission planned to drive around the county to look over possible sites for the nuclear dump. They requested that a county official accompany them on an official "windshield tour." He ended with an open-ended question: "I'm wondering if civil disobedience could stop them?"

Cautious by temperament, I was not so sure the timing was right. "While I'm not opposed to our doing something, I wonder if it isn't premature. After all, they haven't even chosen the finalist sites yet. We don't even know whether one of them will be in our county."

"If we don't resist them now, we can be *sure* the site will be here," Stuart interrupted. "We've got to show them we're tough and will fight."

"Stuart's right," Jim Lucey chimed in. "They won't be expecting it now and we can catch them off guard."

Stuart became more animated. "I think we'll win if we can find fifty people to get arrested. They'll take their dump some place where there's less resistance. But I wonder whether we can find fifty people willing to take this step?"

No one said anything for a moment. Then Gary said, "We can if we've got a good plan and people know that we're not just doing something crazy. Planning is key."

Although I could see that this might be a good political move, I still had doubts. Would we even know their route? "If the siting commission brings in bulldozers and drilling rigs to specific places, we could put our bodies in their path or chain ourselves to the equipment. Then we could plan a disciplined action. This seems dicey to me. I don't like the idea of using cars and trucks in civil disobedience. Someone could get hurt and we'd look pretty foolish in the press. Civil disobedience works because it creates a clear image of injustice and gets sympathy for the victims," I reminded them.

"We all agree," Gary said, looking around the room, "that we can't look like fools. We don't know their plans yet and have to get more information. But if we're going to do anything, we've got to do it now before it's too late."

Stuart had expected that I would be more enthusiastic about using civil disobedience. "When they bring equipment into the county, it'll be too late. I'm not sure we could get people to throw themselves in front of bulldozers. That's a pretty gutsy thing to do. Most people around here haven't even thought about doing civil disobedience at all."

Looking directly at me, Stuart continued, "Tom, you've talked a lot about how disciplined people have to be when doing civil disobedience. If we don't get people together now, we won't have time to train them in nonviolence. Are you willing to proceed to this next step? If the siting commission simply comes into the county and drives around on their own, you're right that we won't be able to do anything. They may have driven around already. But they're making this so-called 'windshield tour' a big deal. They obviously want everyone here to get used to their presence, and they've asked for an official county escort." He paused to let the facts sink in. "That means," he continued, "that they'll have to meet somebody somewhere in the county. We can't just let them waltz in here whenever they want. So what do you think, Tom? You've had more experience in civil disobedience than Gary or me."

"That's pretty persuasive, Stuart," I replied. "As long as we move cautiously I'm ready. In any case you're right that we need to get people trained. I'll tell you one thing, though. There's something wrong with this group of people sitting here. Look around."

"What do you mean? We're all very committed," Gary said, a bit defensively.

"I know what you're thinking," laughed Mark Kurath-Fitzsimmons. "I've been in a lot of protest groups, but have never been in one without any women."

"We certainly can't do this without strong women," Stuart said. "We won't meet again without inviting them to come." Without any further discussion we made this one firm decision.

This group expanded and met regularly for the next five weeks. Half the participants were women. Gary's basement soon became too small to accommodate all the people who wanted to get involved and the meetings moved to a large classroom in Kanakadea Hall, the oldest academic building at Alfred University. "Kanakadea" is an anglicized version of the Seneca Indian name for the geographical region. It probably meant "the place where sky meets the earth," an appropriate image for grounding the many flights of fancy that people took over the next few weeks while discussing possible actions. Hour after hour, in lengthy sessions, no idea seemed too strange. Should we wear costumes to represent the animals of Allegany County when we do civil disobedience? How would you make a red fox costume? Should we rent a plane and drop leaflets over the capital in Albany?

After learning that someone from the county's office of the Soil and Water Conservation District would accompany the siting commission on its tour, the group thought they might trap them inside the building before they got into their car. One night at 2 A.M. Gary Lloyd skulked around the building to get measurements. He came back with elaborate drawings and exact measurements of the building with all its doors and windows.

People explored every detail. How would a group blocking one door communicate with groups blocking doors on the other sides of the building? How would the three scientists who worked there respond? Should we take them into our confidence or was the element of surprise critical? What was the sheriff like? Would he or the state police be in charge of the situation?

John Nesbitt, a technical specialist who worked with ceramicists in the School of Art and Design at Alfred University, finally had enough and told a friend, "Hell would be sitting through these meetings for eternity. I trust everyone here will eventually make good plans, but I'm not coming to another meeting! Just tell me when and where I should go and I'll be there."

On the other hand, Jessie Shefrin, a printmaker and professor of art, found the open exchange of ideas exhilarating, though occasionally tedious. "There's so much creative energy in the group and everyone really listens to each other," she explained as she tried to recruit a friend. "Sure there are a lot of crazy ideas, but it's fascinating to watch the group take them and form them into clever plans. It's the creative process on a group level. We're also learning a lot about each other and building trust."

Sally Campbell's attitude toward the tedium was probably more typical than either John's or Jessie's. A carpenter and artist, she told her husband Stuart, after one particularly tedious discussion, "The only way I can get through these meetings is to remind myself that it's necessary to build consensus. More people are coming to the meetings, and they seem committed. I just wish they didn't have to agonize so much!"

At an early meeting the group got its name—Allegany County Non-Violent Action Group or ACNAG. Jerry Fowler, a local attorney and the assistant public defender in the county, volunteered to represent anyone arrested doing civil disobedience. Although few people noted it at the time, Hope Zaccagni, a painter and graphic designer, began organizing a phone tree that would eventually grow to nearly one thousand names over the next few months; she would soon be spending two to three hours every night trying to keep it current.

In some sessions the group discussed techniques for staying calm during stressful actions. Everyone had to help others stay relaxed. No one should say anything that might provoke the police or the commissioners; silence would be best. A strong support group on the sidelines was essential. Monitors would interpret events for the press and speak for those engaged in civil disobedience; they would also keep other protesters from inadvertently stumbling into the middle of an action. People who were to be arrested would wear orange armbands; the supporters would wear yellow. One person would be designated to communicate with the police to assure them that the group was committed to nonviolence and prepared to be arrested.

❖

MAY 30, 1989—

Jim Lucey learned the details of the "windshield tour" from Fred Sinclair who ran the county office of the Soil and Water Conservation District, and ACNAG called an emergency meeting, attended by about fifty people. He told the group that David Maillie, one of the members of the siting commission, and Jay Dunkleberger, the executive director of the commission, would be coming on the tour. He speculated that Kathleen McMullin might be coming too. He had also learned that they would be driving down from the Rochester airport. They planned to meet members of the county legislature at the courthouse around ten o'clock and then go on their tour. So they wouldn't be going to the conservation office after all.

One woman expressed the general dismay of the group. "It's one thing to block a building with only a couple of doors and a couple of scientists. How're we going to block all the courthouse doors, including the emergency exits? Lots of people are coming and going all the time."

"Before we start talking about strategy," Gary Lloyd interrupted, "I've asked Jerry Fowler to explain the legal details. He's offered to represent us without pay."

Jerry, embarrassed by the sustained applause, stood by the classroom door with his head slightly bowed. He felt he was simply contributing his skills to keeping the dump out of Allegany County like everyone else, and he was committed to ACNAG's general strategy. When the clapping stopped, he walked to the front of the room. "I'm willing to represent you if you're arrested doing civil disobedience with ACNAG, but it's your individual choice whether you want me to."

"Maybe you could explain the legal risks," Gary said.

"Assuming you keep your actions nonviolent, you'd probably be charged with a violation such as disorderly conduct, though theoretically you could be charged with a Class A misdemeanor such as obstructing governmental business. Jim Euken, the district attorney for Allegany County, will ultimately make the decision in consultation with the sheriff's department or state police."

"Could we go to jail?" someone asked.

"It's very unlikely for lots of reasons. Violations are almost always fines. Allegany County's tiny jail is usually overfilled. It would cost the county too much money to put you in jail elsewhere."

"Is there any chance we'd be tried outside the county?"

"Not normally. You'll be tried by a justice in the township where the act was committed. It's possible that a judge might excuse himself because he knows people personally and then a judge from one of the neighboring townships in the county would be brought in."

Hypothetical questions continued. Jerry then added a final word of caution. "I doubt that the charges will be very serious or the fines very large if you cooperate with the police. The district attorney feels very strongly about anyone resisting arrest. When they arrest you, go peacefully. Don't resist arrest in any way."

"If we're blocking a door, a policeman might ask us to leave. Are you saying that we have to do whatever a police officer suggests?"

"No, I'm talking about when they arrest you."

"How will we know when we're arrested?" This innocent question brought smiles to some of the people in the room.

Jerry laughed. "Oh, you'll know. A police officer will say that you're under arrest. At that point you should go peacefully." He paused, then continued, "If there aren't any other questions, I'll leave. It's not appropriate for your attorney to be present should you decide to break the law." Everyone laughed and Jerry left. Even though there were still ambiguities about possible fines and charges, most people felt more comfortable.

Stuart then spoke. "I've been thinking about our plan to block the siting commission in the courthouse. Our old plan presupposed they'd go into the small conservation building. It won't work at the courthouse. I'm thinking it would make more sense to surround their car when it parks at the building."

For the next three hours people discussed logistical questions. "How'll we know where they're going to park?" "How can we prevent them from backing up the car and running over someone?" "Who will be our spokesperson with the sheriff?" "How will we be able to recognize their car?" "How can we maintain the element of surprise?" "If we're standing around won't they suspect we're up to something?"

Someone wondered what we'd do if Maillie said he had to go to the bathroom. Would we let him out of the car? Since we were trying to address injustice, not humiliate people, wouldn't it be against the philosophy of nonviolence to force someone to become so uncomfortable they'd pee their pants? Someone suggested bringing a mayonnaise jar to take care of the situation.

Jim Lucey, finding the discussion ludicrous, left the room for a smoke. "Who cares," he thought, "if they pee all over themselves." As he was leaving he laughed out loud when he heard someone else argue that it would be embarrassing for a woman to have to pee into a mayonnaise jar in a car full of men.

After several more minutes Stuart Campbell finally lost his patience. "No woman in the history of the world has had so much political attention paid to her bladder. We'll bring a jar. Now can we move on!"

The meeting lasted late into the night. Many issues had been resolved and people were too exhausted to reflect any further. Wearing their orange armbands for the first time, they would meet in the county courthouse's parking lot at nine o'clock the next morning and blend in with the peaceful demonstration that CCAC had organized.

❖

MAY 31, 1989 —

Jim Lucey was lurking around the Rochester airport waiting for a flight to arrive from Albany. He didn't want to be recognized by Maillie or Dunkleberger when they got off the plane, so he had tucked his long hair under his cap and was wearing dark glasses. He could identify them, he was sure, because he'd been attending their meetings while he was lobbying legislators in the state capitol. He had even publicly challenged their facts at one meeting. Because of his ongoing political activities in Albany as a key figure in Don't Waste New York (DWNY), a group comprising all ten of the targeted counties, he felt that he should neither face arrest nor publicly identify himself too closely with ACNAG. Having taken part in all the planning, however, he volunteered to identify their car at the airport and report the time of their departure.

According to the monitors the flight was on time, so he decided to call Hope Zaccagni, his contact with ACNAG, and let her know that everything was on schedule. Just as he picked up the receiver, he caught sight of David Maillie walking toward him. "Damn," he thought, "he must have seen me." Instantly he realized that he'd been a fool. "Of course. Maillie wouldn't be flying into Rochester, he lives here and is meeting Dunkleberger's plane." Jim turned his body away from the commissioner on the off chance that he hadn't been recognized. Maillie kept walking toward the phone bank, however, looking rather intense, Jim thought. Maillie picked up the next phone and Jim pretended he was talking.

People at the ACNAG meeting the night before had stressed the need for secrecy. Their plan depended on neither the siting commission nor the police having any advanced warning. They would only have a few seconds to surround the car. To make sure that there would be no leaks, they had even decided not to tip off the Buffalo and Rochester media about the civil disobedience action. The protesters hoped they would be coming into the county to cover CCAC's regular demonstration. Sally Campbell, who had been designated as media spokesperson, did, however, let a couple of sympathetic reporters at local newspapers know what was happening.

Jim continued talking to nobody for a couple minutes after Maillie left to go to the gate. Tension gave way to giddiness as he imagined himself as a character in a B grade private eye movie. Feeling a little foolish and edgy, Jim took a seat and pretended to read a newspaper. He watched Maillie greet Dunkleberger and noted that he was alone. Skulking behind posts, Jim shadowed the two men outside the terminal where he watched them get into a car and drive away.

Rushing back inside, he phoned Hope and quickly relayed the information. "Dunkleberger and Maillie are coming down alone. They'll be arriving in a red car—I think it's a Chevy Camaro. I didn't get close enough to catch the license plate, but it's a rental car and will therefore begin with the letter Z. They should be arriving in Belmont around ten o'clock as scheduled." Jim dashed to his pickup and took a back road, trying to beat the red car to Belmont. He didn't want to miss any of the action if he could help it.

May is a very busy month for farmers in Allegany County. Rich Kelley felt he should stay home and plow one of the many fields on his farm. He and his brother were already behind. Yet he did not want to miss the civil disobedience action at the county seat. He even wanted to be among those arrested. Finally he decided to let the gods speak. Weather reports called for possible thundershowers throughout the night and into the morning. If it rained he would go to Belmont, since he would not be able to plow anyway.

The ground was soaked the next morning when he got up to milk the cows, though the storm clouds were now moving off to the east and Rich knew it would be a beautiful day. Before going to the barn down in the valley, he drove up to a high hill and again wondered why he hadn't built his house there. Looking to the west and seeing endless, unspoiled hilltops rising above the mists in the valleys below, he understood again why he was fighting the nuclear dump. He sensed Mother Earth calling him to witness. "What the hell!" he muttered to himself. "The soggy fields probably won't dry out before afternoon anyway, so I'll head off for Belmont after milking the cows and eating breakfast."

Megan Staffel and Graham Marks lived in a suburb near Detroit, Michigan, with their two small children. He was the head of the ceramics department at Cranbrook School of Art and she was a writer who wrote short stories, essays, and novels. Graham's ceramic sculptures were inspired by the natural world, and Megan's fiction frequently dealt with rural themes. Every summer they breathed a sigh of relief as they retreated from the suburbs to a farmhouse they were renovating in Allegany County. They had learned about the civil disobedience action from their friend Jessie Shefrin and her husband John McQueen, an artist who made sculptures of woven branches. The Marks family packed up their clothes and hastened back just in time to attend the ACNAG strategy session.

The four adults were now traveling in the Markses' van toward Belmont. They talked about their families, their teaching, and their art.

"It's always hard to find time to do my own work when I'm teaching," said Graham, "but ever since Megan and I attended the meeting in Belfast, I've had a hard time keeping my mind on my sculpture. It seems so inconsequential compared to this battle."

"I know what you mean," Jessie replied. "It's been hard to think about anything else. But I feel as if I'm still involved in the same creative process, even though it's now directed toward social concerns and political action." Jessie, like many artists, felt as though she were an outsider to materialistic American culture, where both human beings and nature were commodities to be exploited. Her artistic work frequently explored the "spiritual" processes of integration—connecting people with their inner selves, their communities, and nature.

"You probably don't feel so disconnected, because you're living here," replied Graham. "Neither Megan nor I have ever felt comfortable with the suburban lifestyle we're living in Michigan. Whenever I think about the dump fight, I wish I could be here. It was great being at the ACNAG meeting last night. I felt like I was part of something much bigger than myself."

"I know. It's an organic feeling," Jessie interjected. She was trying to find a metaphor to make her feelings concrete. "From the very beginning I felt I was part of a hive of bees, threatened with destruction. People called everyone they knew. I talked to Stuart and Sally, to a farmer down the road, to Steve Myers, to you and Megan. There was a general buzz that seemed to connect us all. I sometimes think we're part of a larger organism, even though we usually pretend that we're separate individuals. When there's a crisis, we pull together, subconsciously recognizing we're part of something much bigger. That's a kind of spirituality I'm exploring in my drawings and prints."

"I almost began crying at the ACNAG meeting last night," Graham said. "It was such a contrast with our life in Michigan. Everyone there seemed to care about each other. They weren't just caught up in materialistic bullshit. Everyone listened. Sure, people talked too much, but there weren't any power trips. I've never before been in that kind of a group."

"It certainly isn't like a faculty meeting," laughed Jessie. "Are you at all nervous about the action?" she asked as they neared Belmont.

"Not really," Graham replied, "but I'm very excited. I've got so much energy that it's not going to be easy remaining as calm and peaceful as we're supposed to be."

Spike Jones went out to the barn and let his horses loose into the field. It had rained a lot that night so he didn't have to put any water in the troughs. He thought about his old army buddies who had slogged through the swamps with him in Vietnam, and he chuckled, imagining their comments about his getting involved in civil disobedience with a group of peaceniks. I guess I've always wanted to know what it felt like to be on the other side, he thought.

A year before retirement, the army gave Spike three choices for assignment as a recruiter. He chose to come to Allegany County, because he and his wife were both country people who wanted to raise their young daughters in a rural environment. He had planned to retire in Oklahoma and raise horses, but learned that a small dairy farm near Belfast would be sold cheaply to pay off a failed mortgage. When he realized that six Appaloosas were part of the deal, he couldn't resist. Besides, a place with enough water and green grass would cost a fortune in Oklahoma. Every summer he was sure he had made the right decision, but every winter, cold-blooded by nature, he was sure he had not.

Hoping to become a history teacher, he used the GI Bill to get his undergraduate education at Alfred University, where he took courses with Stuart Campbell. When Stuart suggested he get involved in ACNAG, he had flippantly replied, "Aren't you forgetting that I'm a retired army sergeant?"

Rarely one to let anyone else have the last word, Stuart parried, "It's never too late for an old dog like you to learn new tricks. Why don't you come to the meeting tomorrow night in Kanakadea Hall?"

Spike wanted nothing further to do with governmental bureaucracies; he'd had enough of them in the army to last a lifetime. Wanting to be left in peace to raise horses in his retirement, he was incensed that the state would bring nuclear trash into Allegany County to rain on his parade. The cavalier way the siting commission was treating people infuriated him further, reminding him of all the military's bureaucratic bullshit.

Like everyone else, he had gone to the big meeting in Belfast and watched people give reasons why the dump shouldn't be in their county. He had been in the army too long, however, to think that the siting commission was paying any attention. For them, he knew, this was just one more task to be checked off on a score sheet before they brought in the bulldozers. Although furious at the state, he was also disgusted by people who assumed there was a meaningful dialogue taking place between the county and the siting commission.

As he headed off to Belmont in his pickup, Spike thought about why he had joined ACNAG. These people at least were doing something. As a history student he had been fascinated by both the Civil Rights Movement and Gandhi's fight with the British in India. The discussions of principles and tactics at the ACNAG meetings interested him. Intrigued that plans seemed to be made by common agreement, he remained skeptical that consensual decision making would work over the long run. He had learned in the army that those in authority called the shots, though individuals could learn how to use the bureaucracy to their own advantage. Now he would find out whether this was true in all circumstances.

Spike looked at the speedometer, saw that he was going over the speed limit, and slowed down as he approached Belmont. He wondered whether he could stay nonviolent during his arrest. Nonviolent action went against both his nature and his army training. Could he really remain passive if the cops came in swinging billy clubs? He had, however, made some plans. He asked one of the burliest guys at the meeting to keep an eye on him and drag him away from the confrontation if he showed any signs of becoming violent.

"I've never felt more terrified in my life," Carol Burdick told Pam Lakin as she got into her car. C. B., a large, robust woman in her early sixties, was a descendant of the Seventh Day Baptist pioneers who had settled the area. They were a small, but active, Christian denomination, whose actions and ideals reflected the important social and ethical concerns in the nineteenth century—abolitionism, temperance, and women's suffrage. They believed in universal education and founded Alfred University as the second coeducational college in the country after Oberlin, though a couple of other colleges have also claimed that distinction. C. B.'s father had been a professor of biology and dean of the Liberal Arts College at Alfred University when she was growing up.

Although C. B. no longer associated herself with the church, she retained the denomination's strong intolerance to injustice. She had often wished that her life circumstances had given her the chance to participate in the civil rights and antiwar movements, but she'd had to raise three small children alone after a divorce. On one level she now felt a deep gratitude for having the opportunity to fight for the land of her ancestors and against injustice, but she was also very frightened. "Breaking the law is hard for me," she told Pam as they drove toward Belmont. "I'm sure we're doing the right thing, but I didn't sleep much last night and couldn't eat anything this morning. Is there any chance the police will beat us? I don't think I can handle that."

"I don't know the sheriff," said Pam, "but I know his deputy quite well. He was in the police department in Alfred and used to stop in the bookstore where I was working and have coffee in the morning. There's no gentler cop anywhere. Don't worry. The sheriff's department will be cool. And no one in our group will provoke anything."

"Why don't your words make me feel any better? I know you're right, but I still feel like I'm going to throw up." C.B. looked up and saw the concern in her friend's face. "I think I'll be strong enough. But this sure isn't easy. I'd go home right now if I thought I could look myself in the mirror!"

Pam smiled. "Stay with me. We'll link arms together. You'll be okay."

Hope Zaccagni, a small dark-haired woman in her forties, rushed out of her house with the information that Jim Lucey had given her. She knew that it would take the siting commission nearly two hours to drive to Belmont and she could get there in a half hour, but still she couldn't stop her heart from pounding. She forced herself to stop and look carefully both ways as she left her driveway. Her body would feel much better, she was sure, if she could just run to Belmont instead of driving. She was glad that she had written down Jim's words; she'd already forgotten the make of the car, but remembered it was red.

Hope had agreed to call radio stations in the county while the action was taking place. Why on earth did I let Sally Campbell talk me into doing that? she said to herself.

Hope had told Sally that she wasn't sure whether she wanted to be arrested. She remembered Sally's reply. "That's okay. I really want to be arrested and I thought I'd have to make phone calls to the radio stations and talk to the press. Mary Gardner has agreed to talk with the press on the scene." Sally was referring to CCAC's plucky publicity director, who had taken part in all of the ACNAG planning sessions, but could not be arrested and remain CCAC's office manager. "I still haven't found anyone to call the radio stations," Sally continued. "You'd be perfect."

As Hope was driving to Belmont she now realized she'd rather be arrested than have to make those phone calls. "I'm just too excited," she mumbled to herself. "They'll ask me my name and I won't be able to remember it."

Everyone else was sure that Hope was the logical person to call radio stations. She always appeared calm under fire and was extremely sociable. She had been raised in the South and her parents had sent her to a woman's college that served as a girls' finishing school. Although she became an artist and consciously rejected upper-class Southern culture, she never lost her ability to put others at ease. She had opened a small store with another person in the Alfred area in 1968, selling her own jewelry and the pottery of other artists. Eventually she settled in Allegany County permanently.

As she sped toward Belmont she thought about her decision to move here. There were certainly other areas that were equally inspiring in terms of their natural beauty. For Hope the people made the area special. Here you can be exactly who you are and everybody accepts it. Whether you're a doctor or a lawyer or you shovel gravel for the highway department, you're not defined by what you do; you're defined by who you are more than by the job you have.

That's what the siting commission doesn't understand, she thought. At the meeting in Belfast, they only saw a bunch of country folks. They didn't realize they were dealing with intelligent people. As she drove into the village of Belmont, she realized she was willing to be arrested. But she wondered whether she could calm herself down enough to talk coherently to the press.

❖

Every available parking spot, except one near the sheriff's department, was occupied in the lot behind the county courthouse. The Campbells' pickup—containing the all-important mayonnaise jar and blocks of wood for wedging the tires—was parked next to the empty spot. Several times throughout the morning, they had to convince county employees to park elsewhere, without divulging their plan to pin the siting commission in their car.

Right on time, a bright red Camaro drove up to the courthouse. Unexpectedly, however, the car veered away from the visitors' parking area into a public parking lot out front where CCAC demonstrators were chanting. Both Mary Gardner, then publicity director of CCAC, and Glenna Fredrickson, treasurer of CCAC, had taken part in the ACNAG discussions and knew the plan. Mary shouted at Glenna, "We've got to stop them," and rushed in front of the commissioner's vehicle, holding a sign that said "No Nuclear Dump in Allegany County." Other people from the regular demonstration also jumped in front of the car and stopped it dead in its tracks.

"The siting commission's out front!" Stuart shouted. Adrenalin flowing, protesters wearing orange armbands rushed toward the red car as though they were taking part in a chaotic fifty yard dash. In less than a minute they slammed their bodies against the car and linked arms. The CCAC protesters backed off and began chanting "No Dump! No Dump! No Dump!" A supporter wedged wooden blocks under the wheels so no one would get hurt if Maillie's foot slipped off the brake pedal of the running car. Another demonstrator slapped the group's statement face down on the car's windshield announcing that they were "committed to non-violent civil disobedience" and prepared to "fight the commission every step of the way." The statement ended on an optimistic note: "We hope that someday people will see this action as one of many tiny sparks across this land that restored ecological sanity to the earth."

Sheriff Scholes, convinced that the protesters in Allegany County would be peaceful, was startled by the demonstrators' rushing to surround the commissioners' car. Elliott Case, the diminutive owner of Kinfolks Grocery, a small natural foods store in Alfred, had been appointed liaison with the sheriff to explain what was happening. As Larry Scholes rushed toward the siting commission's car Elliott stepped in front of him. "I need to explain what's happening."

Larry, who could easily see that people were linking arms and surrounding the vehicle, did not think the action needed any explanation. He glanced down at Elliott, "I'll talk to you later."

"You don't understand, I'm going to tell you what the protesters are doing," Elliott shouted up at the retreating sheriff.

"Damn!" thought Elliott. "I told everyone last night that it's hard for me to get anyone's attention. It's almost impossible for a dwarf to take a leadership role; everyone expects leaders to be tall." Elliott had been a part of the Alfred Community for more than ten years, and people who knew him felt in the presence of a formidable man and didn't judge him by his physical size.

As planned, everyone surrounding the car remained silent and controlled, though very tense. Drew Robinson, a large and muscular man, was red in the face; he was gripping Spike's arm hard. "Ease up, Drew. You're cutting off my circulation."

"Back away from the car," demanded the sheriff in an authoritarian voice. "We're going to move this vehicle. We're going to allow them to park and do their business." Tension mounted and no one said anything. "What do you people think you're doing?" Off in the background the CCAC demonstrators began singing "We Shall Not Be Moved."

Finally someone broke the silence and told the sheriff, "Elliott Case is our spokesman. You should talk to him."

Larry Scholes finally looked down at Elliott. "Okay. What's going on?"

"These people wearing orange armbands are here to keep the siting commission from doing its windshield tour of Allegany County. They're nonviolent and won't destroy any property or hurt anyone. But they won't move unless you arrest them. The orange armbands mean that they're prepared to be arrested to keep the commissioners from doing their business in Allegany County."

The sheriff turned back to those surrounding the red car. "We don't take law-abiding citizens prisoner in Allegany County. It's my strong opinion that you should take your protest over there and let these people get out of their car. The record you've got in Allegany County has been exceptional so far. Don't spoil it now. You shouldn't do this."

No one moved or said anything. C. B. began to feel empowered by the silent support of people linking arms. The two men in the car had not turned the engine off, however, and she was standing just behind the tail pipe. Now she was feeling nauseous, not from fear but from exhaust fumes.

As the apparent chaos finally organized itself into a carefully planned civil disobedience action, the sheriff began to recognize some of the protesters. Many were highly respected in the community—Bill Coch, a medical doctor and the county health officer, Gary Brown, a building contractor, and Roger Van Horn, owner of a local publishing company. Hoping to lessen the anxiety, Scholes quipped, "I feel like Andy of Mayberry on a bad day," referring to the old television classic, *The Andy Griffith Show.* People laughed. Tension melted.

Just as Larry began to walk away to let everything settle down, several state police cars roared up. Troopers jumped out of their cars and ran toward the protesters, the lead officer drawing out her billy club. The sheriff immediately intercepted her, holding up his hand. "I don't think I'll need your help, but I'd appreciate it if you stayed around just in case. Why don't you wait back in your cars. I don't want to inflame the situation at this point."

About ten minutes later the sheriff returned to the protesters. "What do we need to do to resolve this confrontation?" he asked.

Elliott Case wondered, "Why doesn't he just arrest them?" He again went up to the sheriff and tried to explain. "These people won't let the siting commission out of the car until you arrest them. They've all made a personal commitment. That's why they're wearing the orange armbands."

The sheriff had understood Elliott the first time, but he didn't want to acquiesce to the demonstrators' demands. He hoped there was some other way to resolve the impasse. What would happen if he didn't act? Anyway, he knew it was necessary to let things cool down so no one would panic when the arrests came. He could see that the demonstrators were clearly resolved to stay there all day, so he might as well go prepare for the arrests. He wondered whether he had enough citation forms to arrest this many people.

About forty minutes after the protesters first surrounded the siting commission's red car, the sheriff came back from his office. "I have to tell you that at this point each one of you surrounding the vehicle and wearing an orange armband is under arrest for disorderly conduct. Would you please accompany me to the sheriff's office where we will issue you appearance tickets." Forty-eight people wearing orange armbands followed the sheriff to the holding area and awaited processing.

The county legislators waited patiently in their chambers for the two commissioners. No legislators mentioned the protest; when the commissioners finally arrived, they proceeded as though they were discussing some routine issue of county government. Gary Ostrower, there to report events to Hasper, who was in a legislative session in Albany, hadn't planned on saying anything. After about a half hour, however, he couldn't contain himself. "Doesn't everyone here realize that fifty people in the county just got arrested for civil disobedience? I think it's important that the state of New York realize how strongly people here are opposed to the dump." Commissioner David Maillie brushed him off, saying that such hysteria was unwarranted and irrelevant. None of the county legislators challenged him.

Workers in county courthouse offices, however, had draped signs and banners from their windows protesting the dump and supporting the civil disobedience. Looking up at the windows Stuart Campbell knew that the action had been successful. It had not divided the county, but made nonviolent resistance a respectable alternative to the legal and political battles that CCAC was waging against the dump.

As the protesters emerged from the sheriff's department, smiling and waving their citations, the CCAC demonstrators cheered and whistled. Pam smiled and turned to C. B., "Well, was it worth it?" C. B. just beamed and said nothing.

When Rich Kelly drove back toward his farm shortly after noon, he stopped at a gas station in Angelica. "Did you hear about the people who got arrested in Belmont for trying to stop the siting commission?" the attendant asked.

"No," he lied, not telling him that he had been arrested.

"It was really something. Nearly fifty people got arrested surrounding the siting commission's car. That'll show the bastards that we're serious in Allegany County." Rich knew then that they'd been right to do civil disobedience.

Jim Lucey had been taking pictures of the protest after his return from the Rochester Airport. He thought everything had gone extremely well. Driving home later in the afternoon he turned on the radio and heard commissioner Maillie say that the demonstration "was unfortunate" because it lessened the "possibility of real dialogue." Jim smiled to himself. The siting commission would, of course, like to keep the people in Allegany County talking forever while they built the dump.

The newscast also reported assemblyman John Hasper's reaction. "Although I cannot condone the violation of the law, we must understand two things. . . . First, that the protesters speak for all of us in viewing a radioactive dump as a threat to our lives and livelihood," and second, "that the siting commission can expect precious little cooperation from the people in Allegany County."

"We just hit a home run," Jim Lucey said to himself. He had remained skeptical that Hasper was seriously opposing the dump. But here was a Republican assemblyman who was not distancing himself from the civil disobedience! He was using the occasion, in fact, to emphasize Allegany County's opposition to the nuclear dump. "Far out!" Jim shouted aloud.

3 Declaration of War

Last December, when the siting commission first announced the 10 candidate areas, a local woman said, ". . . They've already picked their site. It's in Allegany County. . . ." I didn't believe her; I swallowed the candidate-area story. I thought there was some reality to the process the state said it was following. . . .

I no longer think so. A hundred things have convinced me that Allegany County was pre-selected. I don't know when, I don't know by whom, I have only a vague idea about how, and I'm only now beginning to understand why. But I'm finding out, and so are a lot of others. Our education is just starting. And none of us will ever be the same.

—Joan Dickenson, reporter for *Olean Times Herald*
September 17, 1989

AUGUST 1, 1989—

"SO YOU'LL COME BAIL ME OUT if I get thrown in jail?" Mary Gardner, the Media Coordinator of CCAC, asked. She was talking on the phone with CCAC treasurer Glenna Fredrickson, but their conversation had nothing to do with finances, money raising, or the media. Tomorrow, Mario Cuomo, governor of New York, would be visiting Alfred University where he would present the New York State College of Ceramics with ten million dollars to build an incubator so that small businesses could develop and market new ceramic products. No governor in the previous thirty years had visited Allegany County, and the anti-dump activists were not going to let the occasion go unchallenged.

"Of course I'll bail you out," Glenna replied. She paused, wondering whether she should again try to dissuade Mary from her proposed midnight adventure. "Don't you want to talk this over with someone else?"

"No. I know you can keep your mouth shut. I can't tell my husband. He already thinks I'm nuts. Since all this started I'm no longer just a stay-at-home mom taking care of kids and cooking meals. I'm changing and he doesn't like it."

"You have changed," Glenna responded. "A couple months ago you wouldn't have told Steve and Jim to shut up like you did at last night's CCAC meeting." She chuckled, "You surprised them so much that they forgot what they were fighting about."

Mary laughed, "They still see me as a ditsy blond housewife. But this lady's finally growing up. Doesn't that sound funny coming from a woman with three young kids?" She continued, not waiting for an answer, "I was Queen of the Maple Syrup Festival in Andover in 1975 and thought I'd done it all. Then I got married right out of high school and started playing wife and mom."

Glenna understood completely. Having grown up in Tennessee, she knew all about domestic roles and social expectations. "Now you've got your own weekly radio show and are quoted in the newspapers daily." Laughing, she added, "'Mary says this; Mary says that.'"

"Isn't it funny that I feel so completely free when the siting commission is strangling our county? Three months ago I'd never think of sneaking into an airport at midnight. It's a rush!"

"You're making it sound better than sex. Why don't I come along and we'll kick some butt together?"

"Steve Myers is going to be angry as hell if I'm caught. If you get caught too, he'll have a heart attack. Besides, who's going to bail me out if you're in jail?"

Isolated from larger and more progressive urban areas and surrounded by farms and forests, the village of Wellsville still retains some nineteenth-century cultural ambiance. Along Main Street at the north end of the village are fine homes (called "mansions" in Allegany County) with manicured lawns, where the remnants of the old oil elite still live. Oil was first discovered south of the New York border in Pennsylvania—oil and gas wells spread into the Wellsville area soon after. The wealthiest people gave money to build a stunning public library, architecturally inspired by the chateaux of the Loire Valley in France. It still defines the center of the village and embodies local intellectual and artistic ideals. An amateur theater company presents plays on the stage of a large auditorium in its basement. Musicians play concerts and artists exhibit their work in a large drawing room just off the library's central foyer, where a huge globe of the world rests. Over the years, mothers and fathers have touched the globe to show their children where they live; now a spot the size of a dime, covering half of western New York, has been rubbed off the world.

Wellsville also contains a large working class. Two factories, Air Preheater and Turbodyne, are located on the outskirts of the city as is Northern Lights Enterprises, a rapidly growing venture that exports candles of wizards and dragons around the world. The marriage of upscale culture with lower-class tastes gives Wellsville its character as a miniature city. Yet the small number of inhabitants ensures that people from all classes meet in social situations; workers' families are just as apt to take their children to the library as are the remnants of the old aristocratic elite who made their money in oil.

Mary, a blonde, youthful-looking woman in her thirties, lived here with her husband, children, and father in the sprawling family home. Her dad had been the Democratic mayor of the village for more than twenty years. Though the county was overwhelmingly Republican, Wellsville was more evenly divided between Republicans and Democrats. With friends and acquaintances from all political and social worlds, Mary's intense involvement with the anti-dump movement gave it both local respectability and instantaneous access to the press.

No one in her household was particularly surprised when Mary told them that she had a speaking engagement and would not be home until late. Returning to Wellsville, she stopped at Tops Market to buy some flour. In the checkout lane she saw Dale Jandreau, an electrician and leader in one of the local CCAC groups. He had just finished stapling Bump the Dump signs to telephone poles along the route from Wellsville to Alfred in preparation for the governor's visit.

"Why're you buying so much flour?" he asked.

"What do you think? We're out of it," Mary replied noncommittally, though her words could not quite hide her excitement.

"Sure, you're buying two twenty-five pound bags of flour, because you suddenly find you're out of it at ten o'clock at night. I know you're up to something. Come on, you can tell me."

Mary didn't need a lot of encouragement. She had so much nervous energy that she wanted to tell someone her plans.

"Why don't I come along and help you?"

"Thanks, but I really want to do this alone. Anyway, I'm going home and wait a couple hours before I go. I don't want anyone to see me." As she walked out of the store, she added, "Remember, don't tell anyone."

When she got home, the rest of the family had already gone to bed. She checked on her kids, who were sleeping soundly, and went to the kitchen to fix herself a bowl of dried cereal and milk. After eating a couple of bites she pushed the bowl away, her stomach churning. She took out a yellow pad and began writing a press statement, but couldn't concentrate. Finally, she slipped out of her house into a shadowy, moonlit world just before midnight.

The local airport, though not far outside the village, was completely isolated, surrounded by open fields. Mary parked her car, turned off the lights, and waited for her eyes to adjust to the dark. Seeing no one, she got out of her car and began lugging the huge bags of flour along a path up a grassy knoll toward the runway. What if a cop stopped her and asked her what she was doing? Light from the moon created an uncanny shimmering world where menacing figures seemed to lurk in the shadows. Staggering under the weight of the flour, she finally made it to the top of the hill, wondering whether she had enough energy to drag the bags any further.

Pausing at the back side of a wooden structure that contained landing lights, she saw a note scrawled on a paper plate: "YOU'RE NOT ALONE, KID." Before she had time to consider the implications of these words, two large dogs bounded toward her, barking furiously. Engulfed by pure terror, she cringed on the ground.

Someone emerging from the shadows jumped between her and the dogs. A man shouted, "Go on home! Get out of here!" She straightened up and was on the verge of running when she understood that the words were not directed at her, but at the dogs. Gradually, her world began to readjust itself; she realized that Dale Jandreau had followed her up the hill. She also sheepishly noted that she had peed her pants and was shaking uncontrollably.

"You insisted that I not come," Dale said apologetically. "So I was just going to stay in the background and make sure you were all right. I hope you're not . . ."

"Believe me, I'm glad to see you!" Mary interrupted. "My legs are so wobbly right now that I'm not sure I could drag this flour any further."

After resting a couple of minutes, Dale carried the two heavy bags toward a flat grassy area near the end of the runway. Suddenly he yelled, "Hit the dirt!" as a small plane buzzed overhead; Mary wondered why a small plane was flying so late at night; she felt as though she had stepped into a Nancy Drew mystery. Was she trying to capture "the smugglers" who were about to land? Before she could figure out her role in the story, Dale brought her back to reality. "Come on. Let's do it and get out of here."

When Mary got up the next morning she was amazed that everyone was acting so ordinary. "This is what it must feel like to lead a double life," she thought. None of her family realized that she had gone out the night before or had any clue regarding her midnight adventure. She had been gone from the house for less than an hour, but it had seemed a lifetime. After her husband left for work, she called Oak Duke, the editor and publisher of the *Wellsville Daily Reporter,* and told him that someone had written NO NUKES in huge white letters at the end of the runway that the governor would be using later that day.

❖

The day before Mario Cuomo came to Alfred the local press reported that Spike Jones would deliver a formal message to the governor on horseback. Conjuring up the image of Paul Revere, Spike would deliver ACNAG's declaration to the governor, asking him to "use all your power . . . to protect the health of the planet and the public, not the interests of private industry." They also wanted to make sure that Cuomo understood the seriousness of their commitment: "Should the siting process continue, we will act as a determined force to protect our liberty, our county, and our lives. We are committed to massive nonviolent civil disobedience and will use all our resources to oppose and stop this madness."

ACNAG had been very open about their plans; Sally Campbell had informed the press. Gary Lloyd had called Sheriff Scholes to explain that Spike would not endanger anyone by riding his horse into the crowd, but would stop at the perimeter. "We've already gotten permission from the governor's office as long as we keep the horse away from Cuomo and the crowd. Would you take ACNAG's message from Spike and carry it to the governor?"

"Why would it be appropriate for me to do that?" asked Larry.

"You could be right there to keep people away from the horse and could get through the crowd easily. No one would worry about someone harming the governor."

"If the governor's office has agreed to the horse, then I have no problem with taking the message."

The morning of Cuomo's visit Spike received a call from Captain Browning, telling him that the state police would not allow the protesters to use a horse to deliver their message to Governor Cuomo. Anyone caught with a horse anywhere near the demonstration would immediately be arrested.

"We've already got permission from the governor's office to do this. We've agreed that the horse will be nowhere near the crowd or the governor."

"The state police won't allow that," Browning reiterated.

"How can you stop me? I wouldn't be breaking any law!" asserted Spike.

"If you're anywhere near Alfred with a horse, you'll be arrested," the Captain responded.

"That'll never hold up in court."

"I don't care whether it'll hold up in court or not. But you're not going to have a horse near any demonstration."

Spike was furious when he called Gary Lloyd. "The governor's office was just playing with us. They weren't ever going to allow us to deliver our message on horseback. We've been so damn naive. It's too late now to make other

plans. I got a horse trailer from a neighbor, but he doesn't want to defy the state police."

"Maybe I can find someone who'll lend us a trailer," responded Gary.

"You'd better find out first what other people think we should do. I don't want to make decisions for the group."

"You told Stuart that you wouldn't wear knickers so that you'd look like Paul Revere," Gary said.

"I wasn't going to wear any goddam knickers," Spike snarled back. "But this shouldn't be my decision. When you talk with the others, tell them I think we should still do it, if we can find another horse trailer."

When Gary checked with other people in ACNAG, few thought they should continue with their plans. The majority thought that the governor's visit, ostensibly having nothing to do with siting a nuclear dump, should not be the occasion for civil disobedience or arrests.

Driving to Alfred later in the morning, Spike realized how stupid they'd been to tell everyone their plans—the press, the sheriff, and even the governor's office. He was especially angry that someone had mentioned his name. Now the state police would be on the lookout for him. He was even more pissed off when Gary told him that no one wanted to challenge the state police.

Ironically, Spike's many years in the army had not stifled his rebelliousness, but nurtured it. "Let me tell you a story, Gary. I was at Fort Dix, New Jersey, waiting to ship off for Vietnam. My wife and I were staying in a hotel. Every day I didn't know whether I'd be back in the hotel room that night.

"They issued us jungle boots and fatigues. The boots had webbing along the sides and drainage holes to pump water out so the canvas dries out fast. This became the 'uniform of the day.' But it's December and freaking cold. There's snow on the ground and they've got us dressed in tropic wear, doing jungle maneuvers in the snow. Now that's stupid! I mean absolutely, totally ridiculous!

"When you're going through training, all ranks are training at the same time. The person training you doesn't necessarily outrank you, but in training they do. There were even a couple of captains going through the training and only a buck sergeant was in charge of all this. Imagine, Gary, we're out in the freezing cold in our lightweight jungle fatigues, with our feet soaking wet in cold snow. A truck pulls up and serves the instructors hot food. Of course *they're* not in jungle fatigues and jungle boots; they're in winter wear and they're eating hot food, while we're eating C-rations out of a can.

"There was a big auditorium there and I noticed that a side door was open. So I said, 'Hey, it's warm in here; let's sit in here and eat our C-rations.' So everybody went into the auditorium, just like a herd of sheep.

"Then, this buck sergeant came in and told us to get out of the auditorium. Well, that was it; I was pissed. I wasn't even a sergeant at the time. But

this was so ridiculous. I stood up and made a speech. 'All of us are here voluntarily, heading off to war, and we know that a quarter of us won't be coming
back. Most of us, with the exception of myself, outrank the people who've
been put in charge of training us. Some of us have been to Vietnam before. I'm
tired of the bullshit of wearing tropical fatigues in a winter environment and
watching you eat hot chow, while I eat out of a can. And by God I'm not taking it any more. Now you get us a warm bus to pick us up, or fuck you, I'm
not going to fight your war.' And I sat back down. There was a big cheer.

"Then this captain came in with two buck sergeants and *ordered* everybody outside. But no one moved. So when I saw that nobody moved, I
looked around amazed, and I stood up again. I repeated, 'We demand a bus.'"
Spike looked at Gary and laughed for the first time that day. "And they went
and got us busses. The next day's training was in the classroom and we never
went out in the snow again.

"So don't you see, Gary, the state police aren't considering us when they
give all these bullshit directions. They just want to avoid hassles for themselves
and stay out of trouble with the governor's office. If we roll over and play
dead now, we'll just seem like patsies. I could hide the horse in the woods behind the university and ride it in. What would they arrest me for? The governor wants to look good in the press, just like the army wanted to look good.
He doesn't want to be upstaged by someone riding in on a white horse, but
he certainly wouldn't want the police to arrest me when I'm doing nothing
wrong. We just have to keep them from arresting me ahead of time."

Gary was persuaded by Spike's passionate argument. If it had been up to
him, he would let Spike openly challenge the state police's edict. "Spike, I basically agree with you. But, dammit, no one else thinks that this is the time to
make our stand. Are you really sure that this is important?"

Spike reconsidered his gut reaction. "I don't know. But I know there's no
end to compromising with bureaucratic bullshit." He paused and Gary waited
for him to continue. "I'm not going to fight everyone else. All I can tell you
is how I feel."

"When I called people," Gary said, "they reminded me that we don't want
to get sidetracked from our main goal—to keep the siting commission off the
land." Gary paused, before continuing, "Remember when a couple of people
wanted all of us to plead Not Guilty after we surrounded the siting commission in Belmont. Nearly everyone decided that we shouldn't get distracted by
fighting the courts or the sheriff when our goal was to keep the siting commission out of our county. A lot of people think this is the same thing."

"Okay. I can accept that," Spike replied coolly. He had said his piece and
it was now important to move on. "We don't have a horse trailer anyway. So,
how're we going to deliver our message to Cuomo?"

Gary said, "I got an old buggy from Wende Bush, the vet in Andover, and a horse's costume. One of us can pull the carriage and the other can sit in it and deliver our manifesto."

"I wouldn't wear any damn knickers and I sure as hell ain't wearing any fake horse's head either."

"You can sit in the carriage, then."

Spike started laughing. "It seems nuts to me, but if the group wants to do this, I'll go along with it."

The state police were on the lookout for Spike and Gary. Lieutenant Mc-Cole found them on a road that circled around the back side of the university. Gary approached the police. "We haven't got any live horses. We're going to draw this buggy with a fake horse and deliver our message to the governor."

Charlie McCole seemed perplexed. Before he could say anything, another trooper interjected, "How do we know that you don't have a horse up there in the woods?"

"You don't. But I'm not going to lie to you. We're going to greet the governor with this carriage, drawn by someone wearing a fake horse's head," Gary replied. "That's all we're going to do."

McCole made his decision. "You can't do that or you're going to be arrested." He dismissed Gary's objections, shrugged his shoulders, and moved away.

Gary walked back to Spike, who was irritated that Gary had initiated another conversation with the state police. "McCole says we can't use the carriage."

"Dammit Gary! Why the hell are we talking to the police? We shouldn't have talked to the authorities in the first place and sure as hell shouldn't be talking to them now."

"They saw us standing here with this carriage," Gary answered. "I wanted to let them know that we didn't have a live horse."

Spike was totally disgusted, as much with himself as with Gary. "I'm not going to have anything more to do with this Mickey Mouse fake horse shit! You and everyone else are broadcasting our plans to the whole goddam world. The police and the governor are making fools of us!"

Though Gary felt Spike was being unfair, he said nothing. ACNAG would have to find another way to deliver its proclamation to the governor.

The Village of Alfred, with nearly six thousand students and only two thousand permanent residents, lies in one of the narrower and higher valleys of Allegany

County. The village's only stoplight is at an intersection on Main Street where a road leads to the two residential colleges, occupying most of the land on opposite sides of the valley. Farther south along Main Street is a block of shops with businesses that mostly cater to students.

Alfred University, founded by Seventh Day Baptists in 1836, is now a nonsectarian private university that contains a SUNY contract college—New York State College of Ceramics—with a world-class school of ceramic engineering and one of the top-rated art schools in the nation. Built on the steep eastern slopes of the valley, the university gives the village a European air. Viewed from the valley floor, buildings seem perched against the cliff like medieval monastic buildings in northern Italy. A president of the university in the nineteenth century spent weekends with his geology students building a stone castle, modeled after one he had seen in Bavaria. Using a variety of large rocks from the odd collection deposited by an ancient glacier, he hoped the building would not only serve functional needs, but that future geology students would study the castle's walls to learn to identify diverse specimens of rocks and minerals. A nearby carillon houses some of the oldest bells in North America.

Alfred State College, an agricultural and technical school in the SUNY system, lies on the gentler western slope of the valley, which had been farmland until the 1940s. Set farther back from Main Street behind the business district and across a creek, its 1950s functional architecture is mostly hidden from view. The college not only serves students across the state, training them in such diverse technical subjects as auto mechanics, surveying, and computer programming, but is also the area's de facto community college.

The governor would first visit Alfred State College before taking his symbolic ten million dollar check across the valley to the SUNY College of Ceramics at Alfred University. At both schools he would encounter antidump activists. He had agreed to meet CCAC leaders during his tour of Alfred State College.

Steve Myers was hoping to persuade Cuomo to challenge the constitutionality of the federal law that forced states to take title to nuclear waste generated by private industries within their borders. He was joined by Sue Beckhorn, who had replaced Rich Kelley as vice president of CCAC, Fleurette Pelletier, action coordinator, and Dave Seeger, their attorney. From CCAC's standpoint the meeting got off to a bad start when the governor decided to hold it in a hallway surrounded by media, rather than in private.

CCAC had organized a petition drive to ask the governor to challenge the constitutionality of the federal law. Elderly women and mothers with children had gone door to door throughout the county, getting signatures from approximately half of all the registered voters. Holding the two-foot-high stack of petitions, Steve handed him the top sheet. "Governor, this symbolizes the nearly twelve thousand signatures I have here."

"I symbolically accept them," quipped the Governor, "but I'm not going to make any decisions based on these numbers. They are minuscule in terms of voters in the state. I've been involved in fights in Queens where we got that many signatures in one evening."

The activists were stunned that Cuomo seemed to belittle their efforts. Wasn't signing petitions to one's representatives part of the democratic process? Didn't he care that CCAC was truly representing the people of the county? When they said nothing, the governor continued. "The point I'm making is that you need to make your case on sound arguments. Even if you had political power, I wouldn't do something just because it's popular."

Dave Seeger jumped into the fray. "The constitutional challenge is based on the Tenth Amendment that the federal government may not abrogate the rights of the states. In this case the federal government is telling the states that they have to take title to nuclear waste, generated by private industry, that has no economic value and is nothing but a liability."

"I'm not impressed by these arguments," Cuomo interrupted. "It's very easy to claim something's unconstitutional, but that doesn't mean it will get a hearing in the Supreme Court."

"I've forwarded a detailed brief to your office."

"When did you send it?"

"About a week ago."

"Three months ago you told me your theory. Why did it take you eleven weeks to get me the brief?" questioned Cuomo. "You can't expect me to deal with this now, can you?"

Fleurette Pelletier was becoming increasingly agitated at the exchange. Although she had been a Democrat all her life and had even run for the county legislature on the Democratic ticket, she found Cuomo's tone demeaning and condescending. Like the others, she had expected the governor to meet them privately for a candid discussion of the issues. Instead, the governor seemed to be using the occasion to posture for the press. "Wouldn't it be better to go inside this classroom?" she asked.

"No, this is fine," Cuomo answered. He was certainly not going to give these leaders a chance to misrepresent him later. They had the right to present their case to him, but this was not a proper occasion for serious negotiations. While he affirmed their right to protest, he deeply resented Republican legislators such as John Hasper who had shifted the focus to the governor's office. "It's your legislative representatives that you have to deal with. I've no power over the siting commission. The legislature voted for the law that set up the commission and is funding them each year."

"But as governor you have to initiate the lawsuit," Steve responded.

"I already told you I'd look into it," the governor snapped.

Steve Myers was undeterred and continued. "There's something else. Your office is writing a certification statement that allows for establishing a *temporary* dump, and that's not acceptable to us."

Cuomo looked puzzled, and Dave Seeger explained, "Your office must certify to the federal government that you will meet the January 1993 deadline for having a nuclear storage facility. The siting commission is currently so far behind that it is doubtful that they can meet it. The language in the certification statement, as now written, would allow the commission to establish a temporary facility at one of the proposed sites before completing the environmental impact statement."

"This is the first time I've heard anything about establishing a temporary site," Cuomo responded.

"It's backed by the nuclear industry," interjected Steve.

"Well, they don't regard me as their favorite governor, you know," Cuomo quipped. He had angered the nuclear industry by helping to scuttle a completed nuclear power station at Shoreham on Long Island. "Let me ask you something," he continued. "Do you think it's wise to put off the decision about storing nuclear waste for future generations, just because you don't want to have it in your community?"

"We're not just concerned about our community, but all of New York State," Sue Beckhorn answered.

"Oh great!" the governor scoffed. "You want to put it in somebody else's state. And they'd say, 'Thank you New York. That's a typical New York thing to do.'"

"But much of the waste coming from the power plants is far too dangerous to store in such a wet environment," Sue countered. "It should be classified as high level waste, and no one suggests that New York is a suitable environment for spent nuclear fuel rods."

"There may be a problem with the classification," admitted Cuomo. "A few days ago at the National Governor's Association meeting, I asked for a new definition for 'low level' waste. That seems reasonable to me."

"That's a little pinhole of light, anyway," Sue commented.

"Let me ask you a question," continued Cuomo. "There's a serious problem of what to do with radioactive medical waste."

"The medical waste accounts for a minuscule amount of radioactivity in the proposed facility," interjected Seeger.

"Don't just ignore the hospitals' arguments because you find them inconvenient," Cuomo interrupted. "If we could change the definition of 'low level' waste to exclude the most dangerous material from the power plants, would Allegany County take the waste?"

"Allegany County would probably not volunteer to accept the waste," Seeger admitted.

"I hear you, and I'm not impressed. You people would do anything to keep the dump out of your county!" The governor had been involved in many battles representing citizens against their government and had mediated one conflict where the City of New York was building a subsidized housing project for low income families in a middle-class Jewish neighborhood. He sought a compromise position and criticized those who took "extreme positions—either for or against the entire project." The last words of the report were those quoted from Edmund Burke, "All government—indeed every human benefit and enjoyment, every virtue and every prudent act—is founded on compromise."

Although he understood that passions ran deep against nuclear waste, he thought that compromise might still be possible and did not see a nuclear waste dump as qualitatively different from other political compromises where one group of people had to sacrifice for the greater good. However, it was only right, he felt, that such a community be compensated for the sacrifice.

"The people in the county are completely justified in mistrusting the siting commission," Steve sputtered, becoming increasingly irritated at the governor's contentious tone. "They won't even give us access to their geological data so that we can check their conclusions."

"The commission ought to bend over backward to honor every Freedom of Information request it gets," the governor conceded, "but you're still ducking the hard questions about what we're going to do with the radioactive waste from hospitals." He paused briefly and then continued, "I didn't create this problem. You should be mad at the United States Congress, not me. I'll look at your constitutionality arguments, but right now I'm not very impressed. These cases take time and even if we initiated it now, it could take up to seven years. Seven years is a long time for a politician. Seven years from now I could be senescent. You're asking for a guarantee, and I can't give you that. We'll take one step at a time." The governor suddenly turned and left the CCAC officers staring after him as he continued his tour.

Cuomo was two hours behind schedule when he approached the plaza in front of the art school across the valley where he would present the symbolic ten million dollar check to President Coll. Police surveillance would make it impossible for Gary and Spike to get close to the governor, so they gave ACNAG's Declaration of Resistance to Stuart Campbell, who would have to figure out some way to get it to the governor.

Security had tightened considerably. "It'll be difficult to get past Cuomo's bodyguards," Stuart remarked.

"Give it to me and I'll take it up," said Megan Staffel. "The bodyguards won't stop a woman carrying a two-year child." She grabbed ACNAG's declaration, bound in florescent orange ribbon, and headed toward the stage. Struggling through the crowd toward the podium, she heard President Coll conclude his welcoming address. "Your visit shows you care about this area. We have tremendous natural resources in Allegany County, and we look to the governor to help us develop *and protect* those assets for future generations." As the crowd erupted in cheers, Megan walked deliberately toward the platform.

When she was a few feet away from the governor, who was approaching the microphone, bodyguards stepped forward and blocked her way. Immediately the crowd started chanting, "Let her speak! Let her speak!" and the governor motioned her forward. She handed him ACNAG's statement as he simultaneously stepped back from the podium. Megan had not planned to say anything, but seizing the moment, turned toward the microphone. Looking around for someone to take her squirming daughter, her eyes met Cuomo's; she smiled and handed him her daughter. The toddler howled the instant he took her and the governor's smile changed to startled consternation. Amidst good-natured laughter, someone shouted, "Maybe the future generation is trying to tell you something, governor." In the back of the crowd, Spike smiled and thought, "This turned out better than a dumb fake horse."

When Sue Beckhorn crossed the valley to Alfred University after her encounter with Cuomo, she saw Spike. "I'm very angry," she sputtered. "Mario just played to the press. He treated us as ignorant yokels. I'm really pissed at his arrogance!"

Spike had never thought much of CCAC's reliance on political and legal tactics. Reasoning with the siting commission to keep the dump out of Allegany County was, he thought, like trying to convince a wolf that it should be a vegetarian. Now he wondered why anyone had expected that an urban, Democratic governor would protect people in a rural, Republican county. Humbled, however, by the state police's checkmate of ACNAG's plan to have Paul Revere confront the governor on a white horse, he felt empathy for Sue, who had prepared her arguments only to have the wolf reaffirm its carnivorous nature.

"Sue, it's time to make plans to keep the siting commission off the land. What do you think about my coming to speak to your CCAC group about civil disobedience?"

"I think it's a good idea," she answered.

The governor's visit was the catalyst for solidifying closer ties between ACNAG and CCAC. Steve and Betsy Myers would henceforth refer to ACNAG as "the underground." Betsy would participate fully in its activities, even while Steve perpetuated the fiction that the two organizations were distinct. Leaders in ACNAG increasingly turned to CCAC for support, especially to help them propagate their message of civil disobedience. In the next few weeks Spike and Gary spoke at community meetings set up by nearly every local CCAC group.

In Allegany County the governor's visit came to symbolize the callousness of urban New Yorkers who cared little about the health and environment of rural people. Sue publicized her anger in a widely quoted "open letter" to "Mario" that ran in all of the local newspapers. She called his attitude "insulting," because, like other "urbanite sophisticates," you treated us "country people" as "ignorant yokels." "What made me angriest," she wrote, "was your remark: 'You people would do anything to keep the dump out of your county,' . . . said in a derogatory manner as if we were somehow narrow minded." The very reason CCAC wanted him to fight the federal mandate, she scolded, was to address the flaws in the legislation, not simply to shift the burden from Allegany County to some other community in the state.

On September 9, five weeks after the governor's visit to Alfred, the siting commission announced "the big losers in a lottery of death," as one activist described it at a rally. Few were surprised that three of the five potential sites were in Allegany County,[1] but there were still reactions of grief. A county legislator whose farm was across a tiny dirt road from the West Almond site tried to convey his feelings to reporters: "If you've ever been hit in the stomach, you know how I feel. It hits you and takes your breath away." Other people expressed their feelings symbolically by wearing black clothing. One woman wrapped a black cloth around a huge locust tree that stood in her front yard.

Very quickly grief gave way to anger and defiance. People in nearly every house in Almond put bright orange "No Dumping" signs in their windows. Assemblyman Hasper declared war at a rally outside the county courthouse and in the press. "I'm telling you there's going to be war out here. . . . This is going

---

1. The other two finalist dump sites were in Cortland County. People also now learned that twenty-nine of fifty-five potential sites were in Allegany County when the Siting Commission first narrowed down the field. Now nine of fourteen "backup" sites were in the county. Since Allegany County's soil composition and other geographical characteristics were similar to much of rural western New York, people found this to be strong circumstantial evidence that the county had been pre-selected for political reasons.

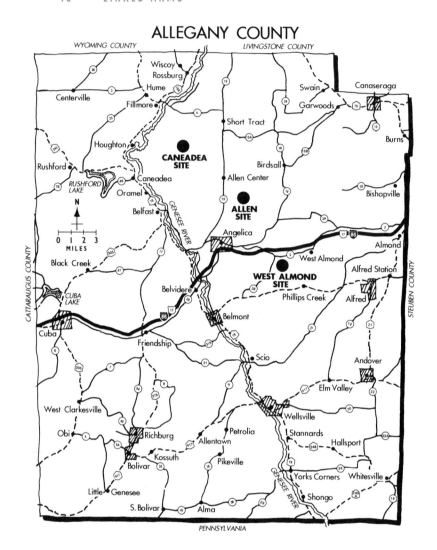

Potential Dump Sites: Allegany County.
Map design by Craig Prophet

---

to be a fight to the finish. . . . The people of this county have risen up and said, 'Enough is enough.'" A few days later, however, he took pains to clarify that he was talking about nonviolent resistance and warned that "any sort of violent act would be devastating" because the opposition "would lose all credibility."

The leaders of CCAC groups throughout the county began championing nonviolent resistance. Steve Myers publicly stated that the fight "will require civil protest" at the sites. Sue Beckhorn declared, "We have to be there as soon as they come into our territory. We have to stop them from doing any testing." Mary Gardner told a *New York Times* reporter, "We'll throw ourselves in front of bulldozers." Glenn Zweygardt said "of course" we'll use civil disobedience, and asked, "How else do you defend your place in the world?" Glenna Fredrickson wanted the county to shut down the roads to keep the siting commission off the land. Gene Hennard warned the siting commission, "If you come into Allegany County, you'll find there's many thousands of people waiting to oppose you. Two hundred years of freedom will not be taken away . . . by your commission."

Ironically, statements from the leaders in ACNAG were more subdued. They took the occasion to encourage people to resist firmly but nonviolently. In private they worried that loose talk of "warfare" and "guns" would frighten people in the county from participating in nonviolent resistance. Sally Campbell emphasized this in interviews with the local press only the day after the commission announced the sites. Violence "would be a grave mistake," she said, for "it would give the state reason for retaliation. We have a big advantage in that we are right—they are trying to poison us. . . . If violence erupts, we are on the same level as they are. Our opposition to violence is unyielding. . . . We are obeying a higher law and a more ancient one—the right to defend ourselves—and we will do it nonviolently."

The citizens' groups stepped up their protest activities in the days and weeks following the announcement. A canoe trip down the Genesee River from Belmont to Rochester dramatized the route that radioactive leaks from the sites would take. Marches rallied people throughout the county. People formed a large Concerned Citizens group in neighboring Steuben County, where Hornell, a city with a population of nearly ten thousand people, was only fifteen miles away from one of the sites. Protesters confronted Cuomo at speaking engagements from Buffalo and Rochester to New York City. Lobbying efforts increased in Albany. ACNAG held more than twenty civil disobedience training sessions at various locations, several attended by over a hundred people.

The day the sites were announced Gary Ostrower formulated a general statement of resistance on behalf of John Hasper. "We will not permit the state to store or emit radioactive trash in Allegany County. . . . We will step up our organized protest. . . . We will resist by all practical means." While the rhetoric was unremarkable, the statement was issued on behalf of the assemblyman, the county legislature, CCAC, and even ACNAG.

Hasper began holding regular meetings in his law office in Belfast to coordinate anti-dump activities. Representatives from CCAC and ACNAG, scientists, legislative leaders, and individuals involved in the lobbying effort in Albany all attended. Although agenda items primarily focused on political and legislative matters, the meetings were remarkable for their diversity. Individuals who were planning to break the law explained their plans to traditional Republican legislators. Scientists suggested ways that lobbyists could use their discoveries. CCAC outlined their legal strategies.

Fred Sinclair, director of the Federal Soil and Water Conservation District in Allegany County, and John Wulforst, a geologist who worked in the same office, were two of the scientists present at Hasper's meetings. They had been working overtime since Allegany County first became a potential site. Their intimate knowledge of the geology of the county made them seriously question the siting commission's conclusions in the original Candidate Area Identification Report. On behalf of the county legislature they had sought to review the original data used by the commission.

When the siting commission stonewalled them, CCAC filed suit under the Freedom of Information Act to get the material. After several weeks of legal maneuvering the Supreme Court of New York agreed that the county had a right to it, and the siting commission handed over two computer tapes, containing the Geographical Information Survey. The tapes, however, were encoded by a privately developed software program, which the commission refused to share with the county, because it did not belong to them but to the private contractor.

Fred Sinclair had already concluded that the siting commission had used faulty data. Learning that the long-sought Geographic Information Survey was unreadable, he angrily told reporters, "We feel like we've been mugged," and, "We were fingered from the beginning."

After the finalist sites were announced, the county had forty-five days for a public response. Everyone at Hasper's first coordinating meeting, from legislators and scientists to the advocates for nonviolent resistance, adamantly agreed that there should be no response. The commission had chosen the sites without addressing issues in the first technical report that scientists in the county had painstakingly prepared. Now the commission had further de-legitimized itself by giving unreadable information to the county.

Fred Sinclair and his staff had mustered considerable technical data that raised serious questions about the ability of the clay soil to retard the migration of radionuclides. Unmapped exploratory gas wells from the early part of the twentieth century punctured the land throughout the area. Fred had gone to Albany and presented this information, but the commissioners ignored it. The commission even disregarded information that the sited areas were on top of a primary aquifer that served a large area of western New York and northern Pennsylvania, though it had excluded other areas in the state for the same reason. The aquifer, the commissioners noted, was not "official," since it was not marked on the maps that they'd been using.

All summer, geologists from the University of Buffalo had been studying a major earthquake fault in the eastern United States that extended from Quebec into western New York. Maps used by the siting commission had conveniently shown the Clarendon Linden Fault to stop at Allegany County's northern border. A couple of days after the final sites were announced, the geologists presented evidence that the fault extended far into Allegany County and possibly even into northern Pennsylvania. They believed it should be carefully mapped, because of its potential impact on gas pipelines and dams.

Although areas in the state close to this fault had earlier been excluded as potential sites, Angelo Orazio, chairman of the siting commission, now publicly claimed that the fault would not create any problem because the "storage facility . . . will be able to withstand any and all perceived or anticipated earthquake activity." Bruce Goodale, the commission's environmental director, admitted that "it may require more careful . . . and substantial design," but declared that it would not exclude the area from consideration.

From this point forward the county would have nothing more to do with the siting commission. The scientists would continue to gather data for an eventual court case. CCAC would challenge the commission in the courts of law. Seeger would represent the landowners on the sites and force the commission to tell them when they were coming on their land. Steve Myers would begin calling the governor's office two and three times a day to get them to delete language in the certification process permitting temporary storage of nuclear waste on any of the potential sites. Jim Lucey, who had become Allegany County's official representative to the statewide organization Don't Waste New York, would begin spending half his time in Albany; he would lobby the legislature to change the law to permit on-site storage at nuclear power plants and to reduce funding for the siting commission. Members of ACNAG would prepare to meet the commission on the site with nonviolent resistance.

❖

NOVEMBER 17, 1989 —

A light snow was falling on a gray, cold landscape as Ted Taylor got into Steve Myers's Impala at 7:30 A.M. for the five-hour trip to Albany. They, along with other activists, had an appointment to meet Governor Cuomo that afternoon. In his early seventies, Ted was one of the preeminent nuclear physicists in the world. Stanislaw Ulam, co-founder of the hydrogen bomb, had worked with

Ted on the Manhattan Project. He called him "one of the very few most impressive and inventive" people he had known in science. Freeman Dyson, who had spent much of his career at the Princeton Institute for Advanced Study and had worked with Ted on the space project, compared him to Einstein in his approach to physics. "Very few people," he added, "have Ted's imagination. Very few people have his courage. He was ten or twenty years ahead of the rest of us. There is something tragic about his life. He was the Columbus who never got to go and discover America. I felt that he—much more than von Braun or anyone else—was the real Columbus of our days. I think he is perhaps the greatest man that I ever knew well. And he is completely unknown."

In one of those ironic twists of fate, Ted had retired in Allegany County in West Clarksville at his wife's family home. In recent years he had been working to rid the world of nuclear weapons and had taken part in many international conferences on disarmament. When he read that five thousand people had attended a meeting in late January to protest the low level radioactive dump, he wondered why everyone was so upset about storing booties and gloves when the world was filled with nuclear warheads, and high level waste was leaking at Hanford, Rocky Flats, and many other places throughout the world.

Earlier in his life Ted might have responded arrogantly and dismissed the protesters, but now he distrusted those who clothed themselves in the mantle of expertise. Maybe it was time to listen and learn.

So Ted began to visit libraries looking at technical reports about the classification of 'low level' nuclear waste and was astonished to find that it ranged from practically harmless material to the extremely toxic. Although the most dangerous "C" class of nuclear waste would account for 0.7 percent of the volume, it contained 93.9 percent of the radioactivity in the proposed dump.[2] This class of waste, coming exclusively from nuclear power plants, included highly irradiated metal parts from the core of reactors, such as the clamps on the nuclear fuel rods.

Disturbed by his discoveries, Ted attended a public meeting in Fillmore on February 17, 1989, called by Amory Houghton, the U.S. Congressman for

---

2. Projections varied over the eighteen months that people in Allegany County were actively fighting the dump. For the sake of consistency in this book I am using the figures from the November 1989 *Source Term Report* by the Siting Commission. These were the same figures that Ted Taylor would later use in the report that he wrote in March 1990, when he argued for on-site storage of "B" and "C" classes of low level waste. These two classes accounted for more than 97 percent of the radioactivity and less than 3% of the volume.

the district. At the meeting Ted was impressed by Steve Myers's comments about the failure of the democratic process to address environmental issues of global import. Steve was incensed that private industry could develop nuclear power without the consent of citizens and now expect people in Allegany County to pay the price. Ted had seen many irresponsible acts by scientists and governmental officials who had knowingly jeopardized the lives of unsuspecting people during the Cold War.

"I was appointed to the presidential commission investigating the disaster at Three Mile Island," Ted told those assembled at the meeting. "It was unbelievable that some of the administrators could be so stupid or lacking in knowledge about nuclear energy. . . . So when someone says, 'We know what we're doing—keep calm,' or, 'You don't know what you're talking about—trust us,' the basis for trust has not been earned by the past performance. And that gulf has gotten even wider since Three Mile Island.

"Frankly, you are right to be suspicious. No one in the world knows how to dispose of large amounts of highly radioactive waste safely for the long term. The past record is abysmal."

A couple of months later Ted joined the fight against the nuclear dump. He attended most of CCAC's steering committee meetings and spoke at rallies, challenging the siting commission. Most significantly, he gave scientific respectability to the movement. When Steve had said that the siting commission was treating people in Allegany County as guinea pigs, a few engineers and scientists at Alfred University deplored his shrill rhetoric and argued for dispassionate scientific objectivity, but took little action themselves. One faculty member in the school of ceramic engineering and two of his graduate students had become prominent "scientific" defenders of the nuclear dump. Now Ted gave CCAC world-class credibility.

Today he was traveling with Steve to present his case to the governor. When they reached Albany after five hours of travel and had lunch, they joined John Hasper and Dave Seeger in a large room in the state capitol. Tables were arranged in a square and spaces were reserved for Cuomo and his staff at the table in front of a painting of Indians "selling" Manhattan to the Dutch. The four men from Allegany County took seats in the middle of the table opposite the governor. Jim Lucey walked in with a couple of members of Don't Waste New York and sat at a side table. People from Cortland County, other antinuclear groups, and a few interested legislators arranged themselves at the tables. The press took up most of the space on one side.

Promptly at 3 P.M. the governor sat down and asked the various groups to identify themselves. He then spoke for twenty minutes. He asked everyone to avoid emotionalism and seek the truth about low level nuclear waste. He reminded everyone that the federal government had thrust this obligation upon the states and that it was unrealistic to expect that congress would revisit the issue.

He announced that he had decided, however, to file suit challenging the constitutionality of the federal law, especially the part requiring states to take title to the waste. Here his presentation was interrupted by vigorous applause. When it ended, he cautioned everyone that the Supreme Court did not often strike down congressional laws as unconstitutional. "I certainly wouldn't have agreed to this suit, however, if I felt the argument was implausible."

As always, he reiterated that the siting commission was an independent entity. Although he had appointed its members in accordance with the New York law that authorized finding a site and building a dump, he had no jurisdiction over it. He estimated that more than half of the other states would make the 1993 deadline and warned that New York State would lose its current prerogative of sending nuclear waste to Barnwell, South Carolina, if his office did not certify that New York was also making progress.

He affirmed his commitment to helping any community that got the facility and emphasized that public health and safety were central. He planned to support a legislative act to provide property owners with equity assurance to guarantee real estate values. He also agreed that the legislature should amend the law to remove all references to "taking title" to the waste. It would make no sense to challenge the federal law on these grounds, if state law required New York to legally own the waste generated by private power plants. Finally, he suggested that everyone focus on the most dangerous "C" class of low level nuclear waste to make sure that its storage would be secure.

The governor paused and waited for a response. The citizens' groups had decided that Ted Taylor would be the first speaker after Cuomo finished his opening statement. The governor's comments about the most dangerous classification of "low level" nuclear waste gave him the perfect opportunity to begin his informal presentation.

"Class B and C waste shouldn't be called 'low level' and shouldn't be part of the proposed facility," Ted began.

"Is it responsible to leave all of this for future generations?" the governor interrupted.

Before Ted could respond, the governor's personal counsel leaned over and whispered something into his ear. Cuomo awkwardly continued, "Now then, Professor, how do you assess the situation?" Hasper looked over at Jim Lucey and both nearly burst out laughing at Ted's instant credibility. They had the same thought—the county finally had someone who could take its case to the governor.

"Class B and C materials have been mislabeled as 'low level.' They come exclusively from nuclear power plants, some right out of the radioactive core.

Every once in a while components malfunction and have to be removed. You can't fix them, because they're too radioactive. How radioactive? If an unshielded cart with a typical package of these radioactive components appeared here, we'd all have a lethal dose within a minute and a half. That's not low level in anyone's definition. That's why so many of us are so vigorously opposed.

"The effort to find a place for the B and C classes of nuclear waste is misplaced. Only 2.6 percent of the volume of material will come from this highly dangerous waste, even though it accounts for over 97 percent of the radioactivity of all the low level waste. The nuclear power plants could easily find space to store this really dangerous material. They're already storing the even more dangerous spent fuel rods. When the federal government designates a place for the high level waste, the same place could accommodate this dangerous class of so-called low level nuclear waste."

The scientist paused, then launched into the irresponsible behavior of those who claimed they could store such wastes safely. "Those who have produced and managed these wastes have a very poor public record. In terms of its capacity to kill, some of this stuff is clear off the human scale. I'm not just talking about the so-called low level nuclear waste, but about nuclear waste management in general. We haven't had any experience with these extremely dangerous radioactive materials on such an industrial scale until the early 1950s.

"In the course of finding out the irresponsible things that happened in Russia, I've discovered equally dangerous things—morally inexcusable things—happening right here in the United States. Radioactive waste was indiscriminately dumped into the Columbia River. That was totally irresponsible and there were people who knew that. There have been many nuclear experts who've strenuously objected to some of the things that we've been doing. There were people at Los Alamos, for example, who were very concerned about what we were doing routinely in Nevada, testing on the surface.

"After tens of billions of dollars, there's no technical solution to permanent, safe disposal of nuclear wastes. There are only more questions. Solving the problems of disposing nuclear waste will take enormous effort and money at the federal level—tens of billions of dollars more."

The governor seemed impressed by Ted Taylor's presentation. "Thank you very much for your participation. We've gained some very important information. I hope that you're willing to work with my staff."

Several others spoke. David Seeger expressed appreciation that the governor would initiate the constitutional lawsuit and offered some technical legal points. Steve Myers was the last to speak. He briefly outlined the many ways that siting the nuclear dump had run roughshod over the democratic process and people's rights of self-determination.

"I think that we may have turned a corner in the fight," Ted commented to Steve on the drive back to Allegany County the next day. "The governor seemed attentive, and his advisor came over afterwards and said he'd like to talk to me later. I gave him my card."

"It was certainly a different story from our meeting in Alfred," Steve responded. "It's a major victory that he's agreed to challenge the constitutionality of the federal law, but that'll probably take too long to stop the siting commission." Steve paused and neither man spoke for awhile.

Steve had read John McPhee's 1974 book, *The Curve of Binding Energy*, and knew that Ted had been a key scientist in designing atomic bombs, including the "Super Oralloy Bomb" and the "Davy Crockett," respectively the largest and smallest atomic bombs ever exploded, and others with such descriptive names as "scorpion" and "Hamlet." At numerous meetings Steve had heard Ted talk about his remorse for having done this. "If I'm not getting too personal, Ted, what made you change your mind about atomic weapons?"

"I don't mind talking about it. It's something I've thought a lot about. There are several answers and I'm not sure I totally understand it myself. In the back of my mind, I never really felt the activity was right.

"Both my wife and mother told me they didn't like what I was doing. Why, they asked, was I making these horrid things whose sole purpose was to kill as many people as possible? I rationalized it like everyone else. 'We at Los Alamos and our counterparts in the Soviet Union are the world's front line of peacemakers. We're making it absolutely clear that war is no longer an acceptable type of human behavior.' They both told me I was talking nonsense, but I didn't want to face the real truth."

Ted paused and became introspective. When he continued, it was almost as if he were talking to himself. "The real truth? I was enthralled at reaching into cosmic energy ten million times more powerful than anyone had ever experienced before. Imagine playing with cosmic power itself! I was fascinated by the possibilities of pressing toward extreme limits in the size, weight, and explosive power of fission bombs. Pushing these limits of performance became an obsession. I couldn't stop pushing until basic physics said the limits couldn't be pushed significantly further.

"I was part of an elite, a select group of people, who were entrusted with dark and important secrets. When I first started working at Los Alamos I was simply fascinated by every bit of information I was given. World famous physicists and mathematicians were just down the hall or up the stairs. Enrico Fermi, John von Neumann, Hans Bethe, Edward Teller, George Gamow, Stan Ulam, Richard Courant, and many others were enthusiastic about what I was doing. They would get excited when I told them how I was reducing the size and weight of fission bombs, while using much smaller amounts of plutonium or uranium-235.

"Within a few months I found that my work was exhilarating, that I was good at it, and that it gave me a sense of personal power over events of global significance. Our work at Los Alamos was strongly encouraged by the President of the United States, the Congress, the entire military establishment, and most of the general public."

Again Ted drifted into silence. Steve waited twenty or thirty seconds, but, fascinated by Ted's story, he finally blurted out the question that seemed to be hanging in the air, "When did you change your mind?"

"When I was working at the Pentagon I suddenly realized the insanity of making these horrible weapons of destruction. When I resigned my job at the Defense Department I vowed that I would use whatever energy I had left to work for the total abolition of nuclear weapons and other weapons of mass destruction. Since that time I've not wavered in my conviction that every nuclear weapon that is made makes the world a more dangerous place."

"But what made you come to this realization?" Steve asked again.

"I've tried to reconstruct exactly what happened during those two years that led to this change in my thinking. In the last several years I've come to the conviction that addiction to nuclear weaponry is a disease. Like true addiction to alcohol or other drugs, the disease is incurable. Some of us are born with it. Like addiction to alcohol or drugs, though incurable, addiction to weapons can be controlled. But total abstinence is the only effective way. To make the vow of abstinence also requires what people call 'bottoming out' so completely that there's no alternative but to admit the addiction publicly and, day by day, to maintain total abstinence.

"Nuclear energy is the same. We're using this unbelievable source of energy to boil water. That's really all a nuclear power plant does is boil water. In the process we're creating a toxic material, having no way to dispose of it."

"I still don't understand what made you bottom out and realize you had an addiction," Steve again inquired.

"I'm not sure I can easily answer that. I guess the short answer is that all of the rationalizations I made suddenly seemed hollow when I was sitting in the basement of the Pentagon drawing circles on maps of Moscow, trying to determine the minimum number of atomic bombs it would take to annihilate all the buildings and people."

"Until now that process seemed so abstract," murmured Steve. "Of course everyone knows that all of our missiles were directed at Russia and theirs at us. Obviously someone was figuring out the logistics. But it's very strange to think of you sitting there and doing that."

"Believe it or not, it seems just as strange to me," Ted replied. "No one can really grasp this madness. I tried to convey the insanity of this arms race to Orlova Galena, who was my guide in Moscow in July 1986, when I was attending an international conference on halting nuclear testing.

"A day before the conference was to begin, I was standing alone halfway between Lenin's mausoleum and the tomb of the unknown soldier in Red Square watching hundreds of happy people milling around. Newlywed couples and their wedding parties were paying homage to their heritage by visiting the two shrines. I thought about the lives of people, both famous and forgotten, who had struggled and died creating the centuries-old Russian civilization. How easy it'd be for a terrorist to destroy the unique architecture and varicolored domes of St. Basil's Cathedral with a well-placed bomb. Creation and destruction: two intense ways for humans to experience the cosmic process itself.

"As I watched people in Red Square, I thought about my own involvement in creating these terrifying bombs and even targeting this very spot. At that moment everyone in the square seemed to be accusing me of future crimes that had not yet happened, and I started crying.

"My guide had left me alone for several minutes while she picked up tickets for some performance. Upon returning, she was shocked to find me in such an emotional state and asked me whether I needed medical attention.

"'No, I'm okay. But I would like to tell you a story, if you have the time.'

"She nodded and steered me over to an outdoor stand that sold cold drinks, and I talked for nearly an hour.

"I told her that I was watching people in the square when my mind flashed back to some work I was doing nearly thirty-six years ago, and I was overwhelmed by the insanity of it all. Imagine—people in the United States, in Russia, and in a few other countries are still engaged in this insanity!

"I explained to her that in November, 1950 I was working at Los Alamos on a much more powerful fission bomb than people had previously thought possible and was sent to Washington, D.C., to spend time talking with our military about their requirements for the new weapons we were developing. I spent one whole day at an intelligence office, poring over aerial photographs of Moscow, placing the sharp point of a compass right on Red Square, and drawing circles to measure the damage a five hundred kiloton bomb would have. I remember being disappointed when none of the circles included all of Moscow."

Ted looked at Steve and exclaimed, "Isn't that unbelievable? I can't really explain all of this, even to myself. I guess that's why I conclude it was an addiction. It certainly wasn't rational."

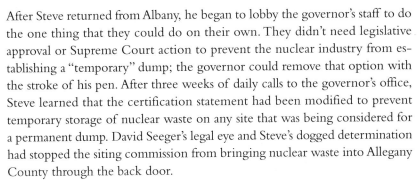

After Steve returned from Albany, he began to lobby the governor's staff to do the one thing that they could do on their own. They didn't need legislative approval or Supreme Court action to prevent the nuclear industry from establishing a "temporary" dump; the governor could remove that option with the stroke of his pen. After three weeks of daily calls to the governor's office, Steve learned that the certification statement had been modified to prevent temporary storage of nuclear waste on any site that was being considered for a permanent dump. David Seeger's legal eye and Steve's dogged determination had stopped the siting commission from bringing nuclear waste into Allegany County through the back door.

ACNAG would have to stop them from coming in the front.

4 Vigilance and Rage

"You'd better think about it and you'd better think again,
'cause Allegany County ain't never giv'n in.
Well now, Allegany County is full of nasty boys.
They fool around with shotguns; they're about their favorite toys.
People say we're nasty, and I say this with a grin,
you ain't seen nothing 'til you haul that poison in. . . .
If you want to bring the poison to put it in our ground,
you'd better bring the army, 'cause you'll have to shoot us down. . . .

—B.A.N.D.I.T.S. song, "Nasty Boys," by Ed and Marta Whitney

SEPTEMBER 24, 1989—

SPIKE JONES LAUGHED, startling his two companions. For a couple of hours he, Gary Lloyd, and Stuart Campbell had been hiking in silence across the largest of the three potential dump sites in Allegany County, zigzagging from one wooded plot to another. They had focused on staying invisible, particularly when they had neared the clear-cut fields of three landowners who seemed eager to sell their property to the state. The stealth of their mission contrasted with the clear, blue sky and warmth of an early autumn day, where the buzzing of crickets mingled with the occasional raucous sound of crows fighting over the rotted remains of a few ears of corn.

Breaking the reflective mood, Spike laughed, "I was just thinking how different this is from slogging through the goddam jungles of 'Nam. God, I hate snakes!"

Startled, Stuart and Gary turned toward Spike as he continued, "I was just thinking how the three of us were sure on different trips in the sixties. Shit! While I was playing Rambo in the jungles, you were doing your professor thing and supporting draft dodgers."

"Draft 'Resisters,' Spike," Stuart interrupted.

"You can't stop playing professor, can you!" After Spike had retired from the military, he returned to college, getting a degree in history at Alfred University, where Stuart had been his mentor. Like most of Campbell's students, Spike deeply respected Stuart's intelligence and usually enjoyed bantering with him. But today he felt annoyed.

"I really don't give a shit what you call them. The point I'm making is that we're three very unlikely 'comrades.'" Spike paused and then took a verbal jab at Stuart. "I'm deliberately using your 'pinko' term so you'll feel more comfortable." Both men laughed, relieving some of the tension that had been building during their day-long skulking behind the properties of landowners hostile to their cause.

Turning to Gary Lloyd, Spike asked, "What were you doing in 1969? I'm obsessed with that year, because it's such a hole in my life. I read everything I can get my hands on about what was going on then in the States."

"I'd just gotten a job teaching high school biology at Alfred-Almond," Gary responded. "That's how I got a military deferment. I thought about joining friends in Rochester who were protesting the war, but my principal was paranoid about student protests and would've canned me. Back then it was unpatriotic to say anything against the war."

"Three different paths, and here we are together in the woods, reconnoitering the enemy," Spike mused. They were now walking back toward the "friendly" side of the site, where the landowners had decided not to cooperate with the siting commission.

After walking a few minutes in silence, Stuart stopped and looked directly at Spike. "This is going to be one hell of a tough site for us to defend. Not only are there seven dirt roads that lead up to the site, there's even that rutted track we first walked on, going right into the center of it."

"Believe it or not, that's actually a town road," Spike interjected. "They don't plow it in the winter, but they maintain it through hunting season for the people who own those cabins."

"They've also got access through the properties of the three farmers who've agreed to let them on the land," Stuart continued. "It'll be a lot harder surrounding drilling rigs if the property owner isn't sympathetic." Even before the official announcement of the finalist sites two weeks earlier, the siting commission staff had been visiting landowners to get permission to come onto their properties.

"Thank God no one on the other sites has signed with the commission," Gary said.

"At least none that we know of," replied Stuart. "I've talked to nearly every property owner on all the sites, but there are a couple of absentee landowners who might've signed."

Shortly after the siting commission's "windshield tour" the previous May, Stuart had obtained maps, showing the routes they had taken. Almost every evening during the summer he traveled those roads, knocking on people's doors. He would introduce himself as a member of the Allegany County Nonviolent Action Group and point to the map saying, "They're coming! This map shows that one of the potential dump sites is right next to your property and may even be on your property itself." Most of the landowners strongly opposed selling their land or having a nuclear dump near them, and they vowed to fight. Many of them would eventually become members of ACNAG.

"Anyway, you've got your work cut out for you, Spike," Stuart concluded. All three men again retreated into silence, wondering how they would defend their sites. Spike thought about the meetings in Gary Lloyd's basement during the previous two weeks when the most active members of ACNAG had gathered to discuss the defense of the sites. When everyone decided that they needed site coordinators, Stuart was an obvious choice. Feeling that he had the strongest rapport with the people near the Allen site, he accepted responsibility there. Gary Lloyd eagerly volunteered to take the West Almond site, only a few miles from his home in Alfred, where he knew several landowners. Spike was reluctant to become the site coordinator at Caneadea, even though he lived nearby. Now, however, heading from a bright sunlit field into the shadows of a darkened woods, he felt at peace with the decision. "What the hell," he thought, "at least nobody's going to take a pot-shot at me here."

As they moved into the woods, Gary's eyes quickly adjusted to the darkness and he saw someone step behind a tree about 250 yards in the distance. "Someone knows we're here," Gary thought to himself, "and we'd better find out who he is." Without saying anything to Spike or Stuart, he wordlessly steered the group in that direction. As they neared the spot, the three men heard a rifle being cocked and saw a man step out of the shadows.

"Who are you? What the hell are you doing out here?" the young man asked bluntly. Stuart thankfully noted that the rifle was pointed toward the ground.

Hoping that he was one of the "friendlies," Spike told him that they were part of the movement to stop the nuclear dump from coming into the county.

The man relaxed, noticing that the three men were wearing orange armbands that emblemized resistance to the dump. "My mother told me you'd visited her. I'm John Clouse." He reached out and shook hands. "You're the guys who got arrested in Belmont, aren't you?" When the men nodded, he added, "Those bastards aren't going to get on my land. That's why I'm out here. Come here, I want to show you something." He quickly led them back into the woods to a clearing where he had been target practicing. "You don't have to worry about them getting on my land," he boasted. "I'll be watching for them; they won't sneak in here!"

A few minutes later the men returned to their cars. "I'm sure as hell glad we weren't the nukers walking out there," Gary said. "That guy was serious."

"I'm not so sure," Spike responded. "Talk's pretty cheap. For that matter we don't really know how many people will do civil disobedience. Frankly I'm skeptical that people will stand up and fight when the time comes."

"It's going to be hard protecting all three sites with civil disobedience," Stuart admitted. "Everyone's got to remain nonviolent. We won't get many people if there's even a whiff of violence. Someone like the Clouse kid could create problems for us. All it would take is one guy to start shooting and we'd lose control. Guns aren't going to win this one. The state's got a shitload more weapons than we do."

"You're right, Stuart," Gary said, "but there may come a time when sabotage becomes necessary and people like the Clouse kid will be the county's last line of defense."

"I'll say one thing for guys with guns, though," Spike said. "They make our position more respectable and may make the siting commission think twice about sneaking on the land."

"Nevertheless," Stuart continued, "we've got to become more serious in training people to do civil disobedience. We'll need at least two or three hundred people willing to be arrested if we're going to protect all three sites."

"What the hell do you think I've been doing while you were out talking to the farmers? All summer I've been going to one CCAC meeting after another, selling civil disobedience. I've been trying to convince people to break the law." Spike paused, then roared with laughter. "If this ain't ironic! A professional soldier selling nonviolence!" Laughing, he doubled over, clutching his ribs.

"Nothing like a convert to become a fanatical preacher," Stuart quipped.

"That's about as far as you'd better take your smartass comments," Spike retorted.

"All kidding aside," continued Stuart, "you can't continue doing this alone and organizing the site, too. It's not enough just to recruit people; we've got to train them in nonviolent techniques. Tom Peterson has done civil disobedience before. Why don't you get together with him and some others."

ACNAG held dozens of civil disobedience training sessions over the next seven months. From the beginning the trainers explicitly articulated four minimal prerequisites for participating in c.d.: making an absolute commitment to nonviolence, assuming personal responsibility, establishing mutual trust, and accepting the legal consequences.

Eventually, twelve people became trainers. During peak periods of activity there were five or six training sessions a week. Three trainers attended each session. The first trainer detailed the history of civil disobedience and explained why it was appropriate for the struggle in Allegany County, the second presented legal issues, and the third discussed scenarios of possible actions and suggested ways to maintain the atmosphere of nonviolence.

The twelve trainers met in a room at Houghton Library and divided up into four teams of three people each. Spike took the lead in explaining how civil disobedience should be presented to people in the county. He had drawn upon his army career as a recruiter and had spoken at many CCAC meetings around the county.

In the early days of the struggle many people had believed that lobbying, scientific studies, and lawsuits could defeat the state. Spike reminded them that the siting commission had completely ignored the county's scientific report and even prevented outside scientists from gaining access to their geological data. He rhetorically asked how people in a small, rural, Republican county could have any political muscle in a large, industrial, Democratic state. He asserted that court battles were both risky and expensive, reminding everyone that the siting commission had chosen the county because it had almost no political or economic clout. "All we have," he reiterated at every meeting, "are our bodies."

Although Spike knew that few people could motivate others as he could, he tried to explain his recruiting strategy to the other trainers. "When you sell something intangible like joining the army, insurance, or civil disobedience, you have to create the idea and let it live. I pick out four or five faces in an audience and speak directly to them. I watch to see how the idea lives in their minds, and I pace myself accordingly. Nothing is worse than talking something to death.

"Remember, you're not trying to convince people to oppose the dump, you want to get them actively involved in nonviolent resistance. Anyone who comes to a training session is already against the dump, but they oppose it for different reasons. I like talking about 'no consent,' because no one knows exactly what it means—or rather it means something different to everyone. For Steve Myers 'no consent' includes the medical risk of exposure to nuclear waste. For Jim Lucey 'no consent' involves the siting commission's arbitrary use of power. Someone else might not like city folks dumping their waste on them; another might be ticked off because they aren't playing honest with the scientific data. A farmer might be pissed because they're planning to take his land. It doesn't matter why people are against the dump.

"You have to give people permission to do what they already want to do—prevent the siting commission from getting on the land. You have to show law-abiding people that civil disobedience is a moral choice so they'll feel okay about breaking the law. There are lots of religious folks around here. That's why I tell them that the Pharoah's daughter broke the law when she rescued Moses. I tell them that Moses broke the law when he encouraged the Israelites to escape. I love history, so it's easy for me to talk about Thoreau. I talk about Rosa Parks and Martin Luther King as heroes. Martin Luther King, for God's sake, even has a national holiday named after him!

"Throughout the whole talk, of course, it's important to emphasize non-violence. For the law-abiding folks I explain that this gives them a higher moral ground. For those who think sabotage and guns are necessary, I explain how and why nonviolence has worked successfully in history.

"If someone is absolutely opposed to breaking the law under any circumstances, however, you're wasting your time trying to convince them. A car salesman can't convince a redneck like me to buy an ugly station wagon, but he can give me permission to buy a pickup truck, because that's what I'd like to have anyway. You've got to give them permission to take action."

Marion Kurath-Fitzsimmons, an assistant pastor of St. Jude's Catholic community in Alfred, first organized and presented the legal information. She told the training group, "All the information I have comes from our lawyer. The legal information gets complicated, because the district attorney has wide latitude in deciding what charges to file. Also there's a broad range of possible fines and jail terms.

"As long as people stay nonviolent, there's no possibility they'll charge anyone with a felony. At the training session in Belfast, a woman asked me if she could be charged with a felony if someone else becomes violent and hits a police officer. The answer is 'no'; each person is only responsible for his or her own action. We're not charged collectively, but individually. We've made it clear that our actions won't entail any violence toward people or property. So if anyone comes to an action with a weapon or tries to provoke violence, he or she isn't part of ACNAG.

"Given all the circumstances, it's most likely that we'll be charged with a 'violation,' like we were in Belmont where the DA charged everyone with disorderly conduct, and Judge Presutti fined us twenty-five dollars, the minimum under state law. There's no criminal record for violations so jury trials aren't permitted. Even though it's unlikely any judge in Allegany County would give the maximum penalty, it's two hundred and fifty dollars and/or fifteen days in jail.

"If they really wanted to throw the book at us, they could charge us with a 'Class A Misdemeanor,' such as 'Obstruction of Governmental Administration.' We'd then have the right to a jury trial, and the judge and jury would be from Allegany County. Misdemeanors do entail criminal records and the maximum fine is one thousand dollars and/or six months in jail. Obviously this is an unlikely scenario, since almost everyone in the county is so strongly opposed to the dump.

"Still, people have to make their own decisions and accept the consequences, so they need to consider the possibilities."

The third member of the training team presented tactics. These varied significantly throughout the struggle. Tactics were always approved by the large ACNAG steering committee. Initially protesters thought that the siting commission would be using drilling rigs, and they would use their bodies to block the roads and keep them off the sites. If they got on, the protesters would link arms and encircle the rig, letting workers out of the circle, but not letting anyone back in. Protesters would refuse to give the police their names, preventing law enforcement officers from handing out citations on the spot; transportation to the county jail would take at least half an hour and slow everything down.

The most important part of the training was explaining why everyone had to keep their emotions in check and maintain discipline. Everyone had to follow the monitors' directives during the action, understanding that there would be no hidden agendas. Those willing to be arrested should wear orange armbands, supporters yellow. Protesters should help each other remain calm and should stand beside people they knew, because trust was essential. Protesters wearing orange armbands should refrain from talking to the siting commission to keep arguments from escalating into anger. To avoid confusion, support people, rather than those being arrested, should talk to the press during confrontations, and only designated spokespeople should negotiate with the police. Above all, everyone had to be dignified.

At the earliest civil disobedience training sessions, trainers conveyed this information verbally. At later sessions there was some role playing. People would stand and link arms, while an "angry landowner" verbally abused them or a "police officer" jostled them. Trainers then suggested that people consider how their emotions might cause an action to become chaotic and violent. If protesters thought they might get angry, they should stand next to calm friends.

All this was fine in theory, but would everyone really be able to stay dignified and nonviolent when the bulldozers and drilling rigs pulled up toward the site?

Three days after announcing the five semifinalist sites in mid-September 1989, the commission announced that "precharacterization studies" would begin in about a month and be completed by Thanksgiving. In December they would select one or two of them to undergo a year-long "characterization study." The protesters therefore expected the state to move rapidly to bring equipment onto the sites.

Most of the affected landowners strongly opposed selling their property or even letting the siting commission onto their land. A grandmother at Allen whose family had owned property since 1817 told the commission, "There is no way you can go on my land." A hog farmer at Caneadea echoed her sentiment: "Until you show me a court order—something from a higher authority—you can stay out." A seasonal resident at West Almond told the commission, "I think it stinks. You can keep your two hundred dollars!"

Jim Lucey and Rich Kelley, promising free legal work, organized a CCAC-sponsored meeting for all landowners for September 22. In the Angelica school gymnasium, attorney David Seeger told landowners that they had certain minimal rights according to the state law that had established the siting process: the commission must notify them when contractors were coming; the contractors must identify themselves before going onto the land; landowners could observe all of the contractors' operations and could get receipts for and duplicate samples of anything taken from their property; and there should be compensation for damages.

When someone asked what they should do if a contractor failed to notify them, Seeger replied that they should contact the police and charge the person with trespassing. Seeger also told the landowners that if they were willing, he would challenge the siting law on the grounds that the state could not temporarily expropriate property to conduct tests that might harm the land without "due process." CCAC would, of course, cover all legal expenses if they decided to challenge the state in court.

Approximately two-thirds of the three dozen property owners joined the lawsuit against the siting commission. On their behalf, Seeger claimed that the precharacterization studies would cause irreparable harm—for example, pristine well water could become contaminated by drilling and some of the planned seismic tests would require cutting down trees. On October 12 he served the siting commission a "summons with complaint," requiring it to answer specific questions in court. A few days later the siting commission retaliated and sought an injunction against the property owners to force access to the sites. After two judges in Allegany County disqualified themselves, both legal actions ended up in Justice Jerome Gorski's State Supreme Court in Buffalo. On November 9 he ruled that the siting commission had the right to go onto the landowners' properties to conduct tests, though they had to notify them in advance and compensate them for any damages.

A few days later, the siting commission announced that it would postpone its investigations for another month, because none of the contractors wanted to be in the woods during deer hunting season. The commission also revised its timetable, now claiming that it would finish the precharacterization studies sometime in the spring. Also, it would not immediately bring drilling rigs and other equipment onto the sites, but would initially send a team to do "limited reconnaissance walkovers."

In the month before Judge Gorski's ruling, however, ACNAG leaders expected the siting commission to begin testing on the properties of landowners who had signed agreements with them, so they created a shadow organization called Allegany County Watch to prevent contractors from sneaking onto the land. ACNAG hung bright green posters with a man holding huge binoculars in nearly every restaurant, bar, grocery store, and service station in the county, asking people to call if they noticed any suspicious activity. Approximately thirty or forty people, many of whom had been arrested in Belmont, became part of a "quick response team," dubbed "the minute men."

Even more ambitiously, ACNAG's steering committee, inspired by the Women's Peace Encampment at the Seneca Army Depot in New York's Finger Lakes region, decided to establish permanent camps near the Caneadea and West Almond sites. (The landowners at Allen were completely unified in opposing the siting commission and told ACNAG that no one would get on that site without their knowledge.) Although Spike Jones moved a trailer onto the Caneadea site, he could not recruit enough people to stay there, and the project soon fizzled.

Gary Lloyd, however, became obsessed with establishing an encampment at West Almond. He got help from a recent college graduate whose family lived only a few miles from the site. Mick Castle's parents had committed themselves to an alternative lifestyle when his father, an engineer, gave up "the rat race of our industrial civilization" after experiencing a heart attack in his early forties. While Mick was growing up, his family built an energy-efficient house, unconnected to any public utilities, isolated in the middle of a woods they called "Pollywogg Holler."

The day the commission announced the sites, Gary contacted the Castles and asked them to drive around the area a couple of times each day, looking for suspicious activities. Two weeks later, Mick joined Gary's outdoor buddies and some Alfred University students who were part of the "World Awareness Coalition" to help establish an encampment. On a blustery Sunday in late September, they built a windbreak, pitched a tent, and moved an old cabover camper to property across from the site. After building a rock foundation for the camper and a campfire pit, they dug a deep hole in the rocky soil, planted a twenty-foot flagpole, and hoisted a huge American flag.

In the late afternoon most of the men drifted off, leaving Gary, Mick, and two students, Pierre LaBarge and Jason Levine, sitting around the campfire.

"Do you guys think that you could spend the night here?" Gary asked. "I hate to leave it unattended just as we got it set up. Could you take charge here for a few days until we can organize some teams of people to take shifts? I've got to teach tomorrow, but I'll come out here right after school."

"I guess I could stay," Mick replied. "I'll go home and get some clothes and stuff."

"I can stay the night with you," Jason added, "but I've got to go to classes tomorrow."

After getting some clothes and other personal items, Jason and Mick rejoined Pierre at the campfire. Munching on potato chips, they speculated about the siting commission's plans. Mick, a wiry-haired, gnome-like young man with a short, bushy beard told the two Alfred University freshmen about all the anti-dump activities that had already taken place in the county. He showed them photographs that he had taken of the civil disobedience action in Belmont, where his father had been arrested.

"It was great last week when nearly everyone in the county turned off their lights and electric appliances for an hour," Jason commented.

"It must have been noticeable at the electric plants," Pierre continued, "but I don't imagine that they'd admit it, even if they keep that sort of data."

"I was in Alfred during the blackout, attending the candlelight vigil in the Seventh Day Baptist Church there," Mick added. "Everything was dark. The Village board in Alfred even had the streetlights turned off for the hour. I really liked that. It unified people in the county and was a positive statement about our willingness to conserve energy."

"That's why I like the idea of this encampment," Pierre said. "We're not only watching for the siting commission, but we're also witnessing to the beauty of the land." He stood up and warmed his hands over the fire. "Hey, why not call this 'the Vigil?'" The name stuck.

Mick stared into the campfire and relaxed. He heard the peepers chirping in the distance and listened to the wind whistling through the trees. He thought about his work at the Cummings Nature Center in Naples, New York, during the previous summer's tourist season, realizing that he liked sharing his love of nature with others. Right now he was glad to be with his two new friends from the city. A few hours later Pierre went back to his dormitory and Mick and Jason crawled into their tent and drifted off to sleep.

A routine was established over the next few days. Jason would go to the university for his classes, take a shower in his dorm, and return to the Vigil with his books. Mick would then go to his parents' house to clean up. Most days someone would bring them dinner or they would rustle up something to eat. When a reporter asked Jason how a serious student could spend so much time at the Vigil, he said that he could study far better in the cramped camper than he could in a noisy dorm. On clear evenings other students would join them around the campfire and sometimes camp overnight.

Usually, Mick was alone during the days, and he spent his time fixing up the encampment—his first project was building a lean-to for firewood and tools. Gary and his friends spent weekends at the camp setting up a citizens band radio and helping with other building projects. One weekend they moved an old garage onto the property; during the next several weeks they put in a wood stove, installed an old sink, made some shelves to create a kitchen area, and built a couple of bunks for sleeping.

The men were not only constructing primitive dwellings at an antinuclear encampment, they were creating their version of a perfect life without the civilizing influence of women. Fulfilling some deep-seated male urge for ritual bonding, these men were attracted to the project for many of the same reasons that other men join with their buddies to go hunting and fishing. But now they could tell their girlfriends and wives that they were protecting the land.

Gradually, Mick became the Vigil's sole caretaker. At first, he thought he would be there for a week or so. But by a month later the encampment had become his entire life. An outdoorsman by temperament, he began creating a life apart from civilization, personally witnessing to the superiority of nature over technology. Whenever newspaper reporters or television crews wanted a visual image to represent Allegany County's fight against the nuclear dump, they would visit the Vigil.

October is a seductive month in upstate New York. Not only does Mother Nature put on a splendid show of autumn color, but she seems to pause from her summer animation and rest for a few weeks before ushering in blustery winter storms. Throughout October, Mick had plenty of company. Sometimes he resented the rowdy Saturday night beer drinking around the campfire that disturbed nature's solitude; other times, he gladly joined in the fun.

He found purpose in the Vigil, helping create the center of Allegany County's resistance. He read articles about nuclear waste and absorbed technical manuals on installing CB radio communications. As the Vigil increasingly became the visible focus for the resistance, Mick and others gathered

West Almond Encampment © Steve Myers 1990; all rights reserved.

publications from major environmental groups about the dangers of nuclear radiation and the irresponsibility of the nuclear industry. A patient and gentle man, he would spend time sharing his perspectives with curious people who drifted into the camp.

Life was idyllic. The days were sunny and warm, the nights crisp and cool. Camping out on the pristine land made the threat of bulldozers and drilling rigs surreal.

Winter, which came early in 1990, beat the siting commission to the encampment. Snow and rain alternated throughout the month of November, and far fewer people visited the Vigil. As the semester neared the end, students became more involved in writing papers, studying for exams, and going home for Thanksgiving and Christmas vacations.

Mick became increasingly isolated. Sometimes he would go several days with only Gary Lloyd paying him regular visits in the afternoons. The CB radio became his primary lifeline to the outside world. ACNAG leaders who were frantically preparing for the siting commission's arrival began to take his presence at the encampment for granted—one less thing to worry about. Sometimes they would raise him on the radio as they sped from one meeting to another on a nearby highway, frequently getting him out of bed just to say hello. Even when they stopped and visited, Mick hid his feelings of detachment, perhaps because he only half consciously recognized the depression that was beginning to consume him. Conversations seldom seemed entirely real; sometimes he felt as though he were watching himself interact with others in a movie. He slept more, ate more junk food, and occasionally found himself drinking too much.

Mick's parents only glimpsed some of these changes, but they began to worry. His father went out to the Vigil, trying to talk him into leaving. "Your mother and I are going to Costa Rica for a couple of months to work on low-technology building projects. We're wondering if it's wise for you to spend so much time alone out here. You can't keep the Vigil going all by yourself."

"Oh, I'm okay. I've got more food than I can possibly eat, and plenty of firewood. Even the sheriff and his deputy brought some food the other day."

"I'm not talking about survival. You're isolated out here. You've put your life on hold. You should get back to your photography or go to work again at Swain with the handicapped skiing program."

"If I weren't here, I'd still be wondering what's going on," Mick responded. "This is my life now. I've worked so hard to create the Vigil. I can't just abandon it. It's become central to the dump fight and it feels right for me to be here now. I don't think anyone else could do the job as well as I can. I'm not sure the Vigil could even survive without me. I don't want to get arrested; I feel that this is *my* contribution to the county's fight."

Mick's father acquiesced. "I can see it's not going to do any good arguing with you. But think about what I've said, and if it gets to be too much, you know where the house is. It's easy to get so caught up in something that you lose all perspective; I know this from my own personal experience."

A quasi-religious fervor had blinded Mick. People dropped by and reinforced his sense of mission. Their praise made him feel important, but it was also a trap, making it more difficult for him to gain perspective on his life; heroes, after all, cannot easily abandon their quests without perceiving themselves as failures.

Spike was one of the few to joke openly around the campfire about the Vigil becoming a shrine. "Maybe we should put earth in little bottles and sell it to all the pilgrims who come up here. We could make buckets of money for the movement. Nothing sells more than religion. The water around here might have some healing properties, and we could charge people to dunk themselves in the pond." Spike sensed that Mick was offended by these jokes. His instinct warned him not to push them too far. A keen observer of human nature, he could tell that his bantering had touched a deep nerve with Mick.

About six months later, in late April after Governor Cuomo suspended technical work on the sites, Spike was one of several people to tell Mick that he had become too involved in the encampment and was no longer needed. Mick had given his soul to the Vigil. He had weathered the rough times. He was the Vigil! How could anyone criticize his commitment?

Feeling deeply betrayed by friends, Mick went home.

OCTOBER 26, 1990—

Certain events help make particular spots of land holy. But land does not automatically become sacred just because something important happens. People make places special by marking them through words and images. Children claimed that they saw the Virgin Mary at Lourdes, but it took priests and merchants, setting up shrines, to create a place on earth for miraculous cures. Soldiers died at Gettysburg, but it was Abraham Lincoln's poetic address that made that particular spot symbolize national sacrifice for the American ideals of liberty and equality.

West Almond may not have been Gettysburg or Lourdes, but flying a huge American flag at the encampment was the first step in creating a special focal point for Allegany County's fight against the nuclear dump. A "Night of Rage," however, made the sited land in West Almond, and by extension the land on the other sites as well, personally significant, even sacred, to a large number of people. A few days before Halloween, a crowd of about 750 people gathered to express their anger at the siting commission and the state. Sue Beckhorn encouraged people to come and express their outrage. She told reporters, "We can promise a night of fire, fury and commitment."

The idea for a "Night of Rage" began as a staged media event. For months members of CCAC had been trying, without much success, to attract the attention of the national media to publicize their struggle. In early October a reporter from the ABC news magazine *Prime Time Live,* called Sue Bechkorn and told her the crew would be interested in coming to Allegany County at the end of October, *if* there were some major event. Sue excitedly announced the news at a county-wide CCAC meeting. "But we need a big event," she said. "We've had rallies, children's marches, canoe trips down the Genesee River, civil disobedience training sessions, quiltmaking, and on and on. I mentioned these things to my contact, but she said they would need something bigger."

"It's too bad we couldn't get another huge crowd to yell at the siting commission, telling them why we won't take their shit," Glenna Fredrickson remarked. "That was dramatic."

"Hey, that's a great idea," said Jim Lucey. "And since the siting commission won't be here, we don't have to worry about violence, and people can really get angry. Let's have a night of rage!"

"Do you think that's wise?" Sue wondered. "We've been trying to promote nonviolence. Wouldn't this send mixed signals? The press might turn against us."

"Nah. They'd eat it up," Jim replied. "Besides people need to vent. There's been too much despair since the sites were announced. Some people are giving up, because they don't think we can beat the state. Even if national TV weren't coming, the county needs a big pep rally; we need to have a night of rage!"

Maybe it'll put a little fear of God into the siting commission, Sue thought to herself. Theatre and drama had always appealed to her. Since the beginning of the dump fight, she had worn a tricorn hat, reminiscent of the Revolutionary War, to all the protests. Out loud she said, "Okay, I'm convinced, but where will we have it? Maybe we could get the gym in Belfast."

"Too formal," interrupted Jim. "It should be an outdoor party—a celebration—something like people going to the county fair. Why not hold it at the Vigil? Most people in the county haven't been out to the sites. If they're going to defend them, they'll need to feel like it's their land."

Mary Gardner immediately saw the dramatic possibilities of a night of rage. Here was something that appealed to her activist instincts. She usually found herself daydreaming through lengthy CCAC meetings with their detailed treasurer's reports, tedious deliberations about lobbying in Albany, and endless discussions of legal maneuvering. "The end of October's great! It'll be just before Halloween. We can light up the night with carved pumpkins and have a huge bonfire. I'd love to work on it with you, Sue."

"If the weather holds up it'll be great, but it's a big gamble," Sue cautiously interjected. "The encampment is in the middle of nowhere. We're going to have to work very hard to publicize it." A practical person, she began to think of all the things they would have to do. "It's not going to be easy. We'll need a big canopy tent that people can get under if it starts raining. I don't even want to think about snow. We'll have to build a stage, and we'll need a big amplification system for speakers and musicians. We have to have lights on the stage, especially for the TV cameras. The nearest power lines must be a mile away, so we'll have to bring in generators. We'll need portable toilets."

"Let's make it a really big event," interrupted Mary. "Get all the musicians around here who've been writing protest music. This can be like Woodstock—hundreds of people out in the open fields."

A Night of Rage captured people's imaginations, and everyone began to add their own details. Rich Kelley and some friends tacked up hundreds of crudely made signs on nearly every tree around the site: "Death to Dumpers!" "This is War!" "Poisoners Beware! We Will Protect our Children!" These messages were interspersed with orange posters of skulls and crossbones and others with no-nuke logos. Musicians under the leadership of Dale Misenheimer, Howard Appell, and Sue Beckhorn formed the B.A.N.D.I.T.S. ("Band Against Nuclear Dumps In This State") that would eventually record three albums, the first of which is now part of the folklore collection in the Smithsonian Museum. Individual musicians wrote several new folk, blues, and rock songs to protest the dump. Glenna Fredrickson carved dozens of jack-o-lanterns that caricatured Mario Cuomo. Mick Castle stacked wood and cleaned up the encampment. Mary Gardner arranged for local speakers and invited Dr. Ernest Sternglass, professor emeritus of radiological physics at the University of Pittsburgh, to present his research about the devastating effects of low level radiation.

For two weeks before the event, Sue Beckhorn spent every waking moment coordinating activities. One day she met with two other musicians, Cher and Howard Appell, who lived in a very rural area at the northern edge of the county, to discuss the B.A.N.D.I.T.S. role at the Night of Rage. Cher, a woman who was raising her six children, ages two to twenty, in a rustic house that she and her husband had built with their own hands, felt that something was missing. "We need a visual focus for our anger."

"What do you suggest?" Sue asked, as she poured her friend a cup of coffee.

"Remember the high school sports rallies where we set fire to the other team's mascot? Why don't we make effigies of the siting commissioners?"

"I actually made one and still have it," Sue answered, sitting down at the table with her friend. "I tarred and feathered it and hung it from a lamp post at the Courthouse the day the sites were announced. I had matches in my pocket and was thinking about torching it, but didn't have the guts with all the cops around."

"We can make some more, if you think it's a good idea," Cher said.

"Let's do it," Sue replied. "We're really pussy cats and wouldn't hurt anyone, but the siting commission doesn't need to know that. Let them wonder about us. I guess the whole night's a huge bluff anyway. I just hope it doesn't really get out of hand."

Almost everything was perfect for a Night of Rage, except that *Prime Time Live* canceled its appearance because of the developing events in eastern Europe; within the week East Germans would breach the Berlin Wall. It was a balmy night for late October in upstate New York. People were comfortable in sweaters as the temperature hovered in the upper forties. The sky was completely clear; stars were visible and a waxing moon cast an eerie light on fields that had been cleared amidst deciduous hardwood forests. A dilapidated farmhouse, deserted years ago, stood in the distance, suggesting that ghosts and goblins might be afoot. In a field across the road from the Vigil spotlights from a Belmont fire truck illuminated full-sized effigies of members of the siting commission, hanging from a scaffold near a large, wooden stage. A macabre Governor Cuomo grinned out from dozens of lighted jack-o-lanterns that marked the perimeter of a makeshift amphitheatre in front of the stage. Blazing bonfires flickered in the distance.

Early in the dump fight Sue and Fred Beckhorn had taken their young girls to every event. They had become involved in the dump fight because of their concern for their children's future, and it seemed appropriate that their children should be part of the protests. Worried, however, that words and images of rage could escalate into violence, they had decided to leave the children at home this night. What if some drunken rowdies started firing guns? What if someone started a fight and the state police came in to break it up?

At the encampment that evening Sue watched a young family pull their tiny children down the road in a red wagon, and she hoped everything would stay peaceful. Turning back to the stage, where she was introducing the speakers and musicians, she heard Gary Barteau, a carpenter from Angelica, singing a ballad about an Allegany County hero who would pick up a gun and kill a driver who was bringing nuclear waste into the county: "The gods, they knew this was not a war of his making. . . . To the gods not a murderer he, but a hero." A few people felt uneasy while others cheered. (This song was never recorded on any of the B.A.N.D.I.T.S. albums.) The crowd roared when Ed and Marta Whitney, a carpenter and nurse, sang that "Allegany County is full of nasty boys. They fool around with shotguns. They're about their favorite toys."

But the night was schizophrenic. Others sang of the wonders of the Genesee River and the loveliness of the land. One singer even sang about the dove of peace that would someday defeat the polluters who had no respect for life. Most speakers talked about the importance of nonviolence, even as they expressed their anger. No speech more greatly reflected the ambivalence of the night than did Sue Beckhorn's. "We have the right to be angry. Anger is a valid emotion," she intoned in a voice that displayed the county's rage. Then she added, almost apologetically, "If anyone in this county is considering violence, let me remind you—you are acting alone." She tried to reconcile inflammatory songs and violent words with the county's commitment to nonviolence. "These actions are symbolic actions that serve a healthy outlet, so that we can get on with the matters at hand."

At the end of the night, however, the demons of Halloween won out over the spirit of reason. Amid shouts, shrieks, and laughter, someone wearing a skeleton's costume torched the effigies of the siting commission. "The commissioners should be here!" someone shouted from the edge of the crowd. A primal energy that had been building throughout this surreal evening in the wilderness released itself and engulfed the area. Then, instantaneously, tranquility returned.

Bob Lonsberry, a reporter who covered the event for the *Rochester Times Union*, wondered whether hatred and anger was replacing the carefully cultivated ideology of nonviolence. Recognizing Gary Lloyd, he asked if ACNAG was now endorsing violence. Gary explained that ACNAG had not organized this event, though it certainly supported the right of people to express their emotions. When it came time to stop the siting commission from getting onto the land, however, "we want to stop this thing with nonviolence. Civil disobedience all the way through." You have to realize, however, that "when the siting commission comes in with testing, we won't have control over everything. We don't control everybody."

The next day Lonsberry began an article that was highly critical of the Night of Rage, by portraying the ambivalence: "There was a troubling mix of signals at a giant outdoor rally in Allegany County last night as more than five hundred people, opposed to a proposed nuclear dump, shared neighborly fellowship with one another and then applauded folk songs with graphic, violent lyrics."

Sally Campbell, the media coordinator of ACNAG, is a small, strong woman with a gift for speaking clearly and concisely. She actually preferred the physical labor of carpentry and the solitude of drawing and painting to the task of cultivating relationships with the media. In the days following the Night of Rage, she repeatedly answered the same questions from reporters: "Did the Night of Rage signal a change in the protesters' plans to be completely nonviolent in confronting the siting commission?" "Would ACNAG really be able to control the situation and prevent people from becoming violent?" "What would ACNAG do if someone showed up at a protest with guns?"

Patiently, Sally explained that ACNAG remained committed to nonviolent resistance. She was confident that members of ACNAG were well organized and well trained. They would ask violent guys to leave the scene; if they refused, ACNAG would report them to the sheriff. With an estimated three hundred people willing to be arrested, she thought that they could keep the siting commission off the land, and their success would in turn prevent violence.

But Sally was no civil disobedience purist. Like the other leaders in ACNAG, she really preferred to call the activity "nonviolent resistance" rather than "civil disobedience." Now she reminded the reporters that nonviolent resistance could be militant. The goal was not simply to witness to injustice and get arrested, but to prevent the siting commission from getting on the land.

The more Sally answered these questions, the more she privately seethed. Why, she wondered, did the press focus on a few violent words at a rally and not on the state-sanctioned violence of subjecting people in the county to the horrors of a nuclear dump?

Sally was acutely aware that the nuclear industry had been lying to people for forty years. They had lied to people in Fernauld, Ohio, while they released radioactive materials into the air and contaminated the ground around a supposed pet food plant. Sally herself had grown up in eastern Oregon, downwind from Washington State's nuclear re-processing plant on the Columbia River at Hanford. Many of her childhood neighbors were suffering from medical problems that had been traced to radiation released at Hanford. "That's real violence," she thought, "not the words in a couple of songs."

Dale Misenheimer, co-founder of the B.A.N.D.I.T.S. and an English professor at Alfred State College, responded to the media criticism. The folksingers, he wrote in a letter to the editor, were like "poets and troubadours." The musicians "simply express the feelings of the people around us." The folks in the county are "ordinary people who feel betrayed by the state." They are rightly angry, and the "fire of rebellion is kindred spirit to the fire that burned in the hearts of those who staged the Boston Tea Party and wrote the Declaration of Independence." He reminded the readers that the distinction between expression and action was deeply rooted in the first amendment of the U.S. Constitution: "We, the musicians of the Southern Tier, feel angry also. But we don't pick up guns; we pick up guitars and sing the truth. . . . We're the cheerleaders of this anti-nuke movement."

The night of rage introduced the B.A.N.D.I.T.S. to Allegany County. At all future events, its musicians would sing their ballads, rallying people to the cause. Whether the county could resist the intrusion of the state without becoming violent would depend on future events. The Night of Rage, however, renewed the people's commitment to protect the sites. Now the land belonged not only to the threatened property owners, but—symbolically, at least—to all the people of the county.

❖

NOVEMBER 15, 1990 —

The uncanny absence of the siting commission puzzled ACNAG leaders who grew increasingly wary as fall gave way to winter. For two months, the "minutemen" had chased phantom siting commissioners all over Allegany County, one time discovering contractors drilling a water well for a farmer, another time finding a geological survey team from the University of Buffalo tracking the Clarendon-Lindon fault.

Only once had anyone encountered representatives of the siting commission. In the first week of November, Drew Robinson, a husky 6' 2" young man, weighing 220 pounds, was patrolling the Allen site when a farmer tipped him off that two men from the siting commission had tried to get permission to come onto his land; they were driving a red car. He raced off in the direction the farmer indicated. Within a couple of minutes he saw the car parked on the edge of the road, its engine running. He pulled his pickup across its path and saw two men inside looking at a map. The driver lowered his window a bit and Drew asked them if they were from the siting commission. The two men looked uneasily at one another. Drew made a grab for the keys that were in the ignition, but the driver slammed the car into reverse, turned, and tore off in the opposite direction, leaving Drew standing in the road rubbing a bruised elbow.

For more than a month Mick Castle had left the encampment for only short periods of time to take a shower and wash his clothes. Even when Peg Jefferds, a homemaker who was involved with ACNAG, volunteered to spend wednesdays at the Vigil, Mick would only leave for a couple of hours and then return to install radio equipment or work on some other project. On an interpersonal level Peg was not forceful, though paradoxically she was a tiger in her willingness to stand ground against the siting commission.

Peg was at the encampment on November 15 when several men came to do carpentry work on the old garage. One of the men confronted Mick. "Why don't you take the day off. Peg's here and we'll be here all day and can call someone if there's anything suspicious." Mick had no idea what he would do, but he thought it might be good to get away. What happened later, however, would further reinforce his conviction that he shouldn't delegate his responsibility at the encampment to others.

Early in the afternoon one of the workers at the encampment saw a red pickup drive onto the property of an absentee landowner across from the encampment. He watched as two men got out of the truck and started carrying tools and other materials toward the center of the site. Heart pounding, the man rushed into the garage, picked up a recently installed phone, and called to activate ACNAG's emergency phone tree. Within half an hour two dozen people arrived, including Sally Campbell and Spike Jones. Several cars pulled up alongside the vehicle to block it from leaving.

Kathryn Ross, a reporter for the *Wellsville Daily News*, and Sheriff Scholes arrived on the scene just as Sally was organizing the resisters near the red pickup. Kathryn turned to Larry. "What's going on?"

"I don't know," Larry answered. "The siting commission didn't tell me they were coming to do tests. I thought they said they weren't coming until after hunting season."

"What will you do if the protesters start harassing some contractors?" Kathryn asked.

"Right now, I'm observing. I don't know what's going to happen, but maybe I'll learn something," Larry replied.

Kathryn drifted off toward the group of protesters and heard Sally saying, "Remember to remain calm when you confront the siting commission, but surround them and keep them from doing their tests." The reporter quickly counted twenty-four protesters and noted that Spike had just handcuffed himself to the truck's door handle. As the rest of the protesters started to go across the field, she turned to Spike. "Why are you handcuffing yourself to the truck?"

"This'll keep them from leaving." Spike paused, then said, "Shit, I forgot to give the keys to someone. Here, take them and give them to Sally or someone when you get the chance."

"I guess I can do that," said Kathryn, taking the keys from Spike. "You know, don't you, that now you're my prisoner," she said smiling as she followed the group toward the woods.

Just as the ACNAG group neared the woods, two men came toward them. "What're all you guys doing here?" one of them asked.

"That's just what we were going to ask you," replied Sally.

"My brother owns this land; me and my friend are building a deer stand, getting ready for hunting season."

Quickly Sally apologized and explained that they were trying to stop the siting commission from building a dump on the land. The protesters left, heading back toward the encampment, and the reporter remained with the two men. "What do you think of the state wanting to build a nuclear dump on this site?" the reporter asked.

"Well, I like hunting in the area, but I'm not too concerned about a nuclear dump, 'cause I don't live around here."

Kathryn asked a couple more questions then started back toward the encampment to get reactions from the demonstrators. One of the protesters came running up to her, "Spike wants me to get the key from you."

"I want to get a picture first," Kathryn responded. "Then I'll give him his key."

As she walked up to the truck, Spike was yelling, "Give me the damn key."

"I'm going to take a picture first."

"Come on, don't do that. I already feel like a raving idiot for 'cuffing myself to some deer hunters' truck. Don't make me look like a blaming fool to everyone in the county."

"Don't worry, Spike. I won't use your name," the reporter answered.

"Shit! You know everybody'll know me if you take my picture."

"Relax, Spike. What'll you give me not to splash a picture of you across the front page of the newspaper?" she kidded.

"Dammit! Just give me the goddam key!"

"All right, I'll just take a picture of your hand, cuffed to the handle." Quickly she took the picture and handed him the key. As he struggled with the handcuffs, Kathryn laughed and headed toward the encampment. True to her word, she published the picture without verbally identifying him. The caption read: "Handcuffing themselves to vehicles . . . is just one of the ways ACNAG members plan to tie up the commission's study of nuke sites in the county." The edge of Spike's distinctive, fur-lined denim jacket, however, was in the picture, and folks razzed him for weeks.

When Mick Castle returned, he believed that the whole fiasco would have been prevented had he been at the encampment; he would have checked out the people before activating the phone tree. He also worried that it could have been a disaster if the hunters had been more belligerent. Everyone else, however, thought that it had been a good training exercise and were pleased that the quick response team had gotten to the site so fast. Over the next hour dozens more people showed up at the site, and the tension gave way to a party.

This, however, would be the last "false alarm." The next time protesters gathered on a site they would confront the siting commission itself.

5 Defiance in the Bitter Cold

> The battle between the nuclear industry and the county will now take place at the sites in the cold of an upstate New York winter. This will be our Valley Forge.
>
> —Sally Campbell, Media Coordinator for ACNAG, quoted in the *Wellsville Daily Reporter*, December 8, 1989

A FEW FARMERS GOT LETTERS informing them that the siting commission would be sending technical teams onto their land on December 13, 1989. The commission notified the sheriff that they expected his help in getting onto the site; five days earlier a delegation from the state police had informed him that he could expect no help.[1]

In every interview with reporters from county newspapers and radio stations, Sally Campbell encouraged ordinary folks to come to the site to protest the activities of the siting commission. If they did not want to be arrested, they could peacefully demonstrate and support those who did.

Two days before the technical team would arrive to begin its "precharacterization" study, ACNAG placed a half-page advertisement in the three daily newspapers. Forty-two prominent citizens had signed "A Statement to the People of Allegany County and Western New York," asking citizens to "drop whatever you are doing, gather your friends and family, go to the site and peacefully demonstrate your opposition to this process which would sacrifice Allegany County, its citizens and children to the nuclear power industry." People should "take inspiration from the recent example of the East Germans, who have courageously taken possession of their destiny" and "walk across the site, giving witness to your identification with this land and its people." Among

---

1. The encounter between the state police from Albany and the sheriff is recounted in the Prologue.

the signatories were eleven medical doctors, twenty-one religious leaders, the president and provost of Houghton College, the provost of Alfred University and the dean of its Liberal Arts College.

The ACNAG leaders were not entirely sure how they would stop the siting commission from doing their walkovers. The basic plan was simple. Protesters would link arms and surround the technical team before it got on the site to prevent it from conducting its precharacterization study. If this proved impossible, they would surround the team on the site. Privately, leaders of ACNAG were worried. They were not sure that there were more than one hundred people actually willing to be arrested, and they had to defend three sites, not just one. The site in West Almond, close to the two colleges in Alfred, was about twenty-five miles southwest of the Caneadea site, which was near Houghton College. The Allen site was about halfway between the two.

The protesters had no clue about the siting commission's destination. In planning their strategy they relied on the recent court order, which allowed the technical teams to enter the properties of the resisting landowners, but also mandated the siting commission to inform them in advance. Jay Dunkleberger, executive director of the siting commission, had sent landowners a six-page letter on December 2 that explained what the technical team hoped to accomplish. It would "spend several days at each site," though "the exact amount of time will depend on the weather." The purpose of these walkovers was to study the geology and hydrology of the area, to take shallow soil samples, and to map the bedrock. They would study the surface flow of water and observe erosion features and slope stability. They would evaluate the habitats of both terrestrial and aquatic species. They would also determine whether any land was culturally important to Native Americans.

In the same letter Dunkleberger promised to inform the landowners in advance "what days we plan to be on your property." Five days before the arrival of the technical team, the siting commission reiterated to the press its desire for good relations with the landowners. On December 8 Dunkleberger told a newspaper reporter that they would "like to have landowners walk over the land with us to identify their concerns."

ACNAG knew of only two farmers who had received letters from the siting commission, informing them that they were coming on December 13 to begin the precharacterization studies of their land—William and Lois Clouse at the Caneadea site and Donald and Gloria Miller at the Allen site. ACNAG strategists thought it likely that the three farmers at Caneadea who had already accepted two hundred dollars from the siting commission and one absentee landowner at West Almond had also received letters.

Stuart Campbell was certain, however, that at Allen only the Millers had received a letter from the siting commission. Landowners who lived there were well organized; they had contacted the site's six absentee landowners and knew that none of them had received any notice that the technical team was coming.

Because the court order required the siting commission to notify landowners in advance, members of ACNAG were fairly certain that the technical team would enter the Allen site on the Miller's property, if they came to that site at all. Since the Allen site stood between West Almond and Caneadea, ACNAG strategists also thought they could move people there rapidly. They therefore deployed a small but committed group at the Miller home. Together with Don and Gloria Miller, ten other people surreptitiously waited in the house. Drew Robinson patrolled the back roads and drove around the site. A stronger contingent of about forty people were stationed in various places around the West Almond site. Uncertain about their numbers and particularly worried about defending the Caneadea site, ACNAG asked everyone else to go to the small German Church that stood on its borders. It was also to be the meeting place for the press.

DECEMBER 13, 1989, MORNING—

At 6 A.M. the first protesters began parking along Shongo Road across from the Old German Church at the Caneadea site. Temperatures had plummeted to below zero during the night. Huge clouds of vapor from running trucks and cars billowed into the pre-dawn, starlit sky. Farmers, used to getting up early, hustled about outside the church setting up an outdoor soup kitchen, while their wives served hot chocolate and coffee to groggy protesters huddled inside their cars. People slipped in and out of each other's frost-encrusted vehicles to exchange greetings. The twenty or thirty early birds repeated the same question, knowing full well that no one had the answer. "When do you think they'll show up?" Like odds makers in Las Vegas, people argued probabilities. Most people thought it wouldn't be for a few hours, since city folk aren't used to getting up very early in the morning. A few others thought that they would come within the next half hour to take them by surprise before the bulk of protesters arrived.

When the sky lightened, shadowy trees became part of a crystalline fairyland as light sparkled on ice-encrusted limbs. Winters in Allegany County are long and snowy. Since a blanket of clouds usually insulates the land, daytime winter temperatures rarely fall below fifteen degrees Fahrenheit. Today would be an exception—cloudless, windless, and very cold. The protesters shook off their remaining grogginess, turned off their engines, and entered into the frigid beauty. For a little while, at least, the enchantment of the empty countryside, dotted by a few farmhouses along the road, released tensions that had been building during the week. The members of ACNAG had made all the preparations they could. Now the fates would rule.

Gradually, a bustle of activity began to shatter the stillness of dawn. A pair of cardinals and a flock of purple finches tussled and fed at a bird feeder beside a nearby house. Pickup trucks and cars streamed into the area. Sounds from a bullhorn drew folks to the tiny church.

Sally Campbell was using the bullhorn to rally the troops and give them last minute instructions. "Keep your eyes open. If you see the sheriff's big, white car, do what you're doing: stand in the middle of the road. If the technical team begins to get out of their cars, surround them, so they don't get on the land. If they do get on the land, track them down and surround them so they can't do their work. Reinforcements will come from the other sites as soon as possible.

"Remember to remain nonviolent. If any of you haven't been to one of the civil disobedience training sessions, see one of the monitors wearing a white sash. One final instruction for those planning to be arrested. Don't give the police your name. Say that your name is Allegany County. Give driver's licenses, wallets, and anything that can identify you to supporters who are not planning to be arrested."

Someone in the front of the crowd asked, "Why aren't we supposed to give them our names?"

"We want to slow the process down and keep the siting commission from doing its job," Sally explained. "If we give the police our names right away, they might just issue us appearance tickets for court. If we don't give them our names, they'll have to transport us to Belmont, a few at a time, and that'll slow everything down."

"How about going limp when arrested to slow the process down even more?" someone else asked.

"That's a possibility if you want to do it. We're not recommending that now, because it could possibly lead to an additional charge of resisting arrest. We also need to remember that we're not trying to hassle the police and should be as dignified as possible for the press. But if you want to go limp, that's your personal choice. Just think carefully about the possible consequences."

Spike Jones went into the middle of the group. As always, he was wearing a cowboy hat and boots. His jean jacket covered a couple of layers of sweatshirts. When someone asked him why he wasn't wearing a warm coat, he answered that he couldn't afford one. The more probable reason, however, was that bundling up didn't fit well with his cowboy image. At one point that morning, to the delight of the television reporters, he rode his Appaloosa across the snow-covered landscape, checking out the site. That image even made the CBS *Morning News.*

Sally tried to hand him the bullhorn, but he waved it aside. "Y'all can hear me without that thing, can't ya? I'm going to be sending groups of people to different parts of the site, but I want most of you to stay here until we find out where the siting commission decides to go.

"A few minutes ago five guys rushed toward their cars. I asked them where they were going. They said they heard that the siting commission was coming up East Hill Road and they were going to meet them. There was nothing to it. It was just a rumor. We can't react to rumors. We can't start flying all over the place. I don't want any of you driving around the site unless you have a good reason. Unless a monitor tells you to go, stay put."

Spike's teeth were chattering. "None of you are cold, are you?"

One man shouted, "Hell no! This feels like Florida." Others joined in the good-natured bantering.

Spike laughed and said, "Well, if any of you get cold, the Muras have opened up their house to us. Jan is out driving the school bus and her husband will be in the barn most of the morning, so just walk in. She's got a huge pot of soup on the stove if you get hungry. Someone's there now on a CB radio. It's just a half mile walk down this road on the right. We may be waiting here for quite a while."

As the morning wore on and there was no sign of the siting commission at any of the sites, the protesters became increasingly frustrated. The siting commission had said that they were going to spend several days on each site, and this day seemed half over for those who had gotten out of bed at 5 A.M. Rumors that placed the siting commission in various locations were spreading. At 10 A.M. fifteen people in several cars roared up to the German Church. Half the people from the West Almond site, driving faster than they should, had responded, because someone had made a telephone call and told them that the technical team was at Caneadea. Gary Lloyd and the others, chastened and worried about leaving West Almond so marginally defended, rushed back to their site. The lines of communication quickly became formalized, with only a few people authorized to confirm the presence of the siting commission via CB radio or telephone.

Stuart Campbell had asked Don and Gloria Miller at the Allen site if the protesters could wait for the technical team at their house. Incensed that the state could barge onto his property without his permission, Don agreed without a moment's hesitation. Although personally more inclined to pick up his shotgun, he was impressed with ACNAG's resolve and wanted to see if nonviolent resistance could work.

Knowing that there would be few people stationed at the Millers' house, Stuart selected the group carefully. Four local neighbors of the Millers were logical choices. They had all been to civil disobedience training sessions and had participated in lengthy ACNAG steering committee meetings to discuss strategy. The four local resisters were joined by six other members of ACNAG who had decided to face arrest.

At an earlier strategy session someone suggested that a few people camp out on the sites in case the technical team tried to sneak on the land during the night. There was already an encampment at West Almond and Mick Castle and others maintained round-the-clock surveillance. Although most people at the ACNAG strategy session felt that this precaution was unnecessary, the coordinators at Allen and Caneadea agreed to find a few people to spend the night on their sites. So Stuart Campbell asked Don Miller whether it would be possible for a couple of people to spend the night at his farm. Again, Don immediately agreed.

Two women, one in her late forties and the other in her early sixties, drove out to the farm at 5 P.M. on December 12 to meet the Millers. Both were faculty at Alfred University. Pam Lakin, the younger woman, was a reference librarian and Carol Burdick (usually called "C. B.") taught English. Pam had participated in marches in Montgomery, Alabama, during the Civil Rights Movement and understood the potential dangers of civil disobedience. Regretting that she had not become more active in the protest movements of the 1960s, C. B. had the fervor of a religious convert, although her stomach churned uneasily when she imagined the confrontational aspects of civil disobedience.

Both women agreed to spend the night at a farmer's house on the Allen site, when Stuart asked them. Both had been at the ACNAG meeting and knew the reasons for their presence. Imagining that their Allen quarters would resemble the rustic nature of the West Almond encampment, they gathered up their sleeping bags, expecting to sacrifice personal comfort for a bed of straw in the loft of some farmer's barn.

After following a rough map that Stuart had drawn, they arrived at an isolated house in the middle of a moonlit frozen world. Christmas lights filled nearly all the trees and bushes and blinked all around an exterior deck. A special contraption raised lights in increasingly smaller concentric circles up a huge flagpole, highlighting an enormous American flag perched at the top. And, of course, Santa had already harnessed his reindeer to a sleigh in the front yard.

The house itself was a large two-story, ranch style structure, with wings extending out on both sides, where the married son and daughter had their own quarters. The two dazed women stepped into a living room, magically transformed by twinkling lights, tinkling bells, candy canes, dancing dolls, frolicking bears, and especially Santa Clauses. Every surface was covered with Christmas. Everything was immaculately clean and nothing was out of place, not even a newspaper or magazine.

For a moment, as they greeted Don and Gloria Miller, the two women almost forgot why they were there. Santa's magical kingdom and the looming confrontation with the siting commission could not be simultaneously merged into a single thought. After showing them around the more public rooms of the house and exchanging pleasantries, Gloria Miller steered them up a stairway to a bedroom on the second floor where they no longer expected a bed of straw. They would spend the night in a king-size bed with freshly washed, lightly scented sheets. Pam wondered how they would know if the technical team did sneak onto the site.

For a moment, both women were alone with their thoughts. C. B. thought about one of her ancestors, Abigail Allen, wife of a president of Alfred University in the late nineteenth century. C. B., a nationally published essayist, had recently written a one-act monologue with Abigail as the sole character. In the play, which C. B. had performed in costume on several occasions, Abigail explained her reasons for defying the law. In 1887 she had become nationally famous (or notorious) when she marched into a polling place with nine women friends. They grabbed ballots off the table, marked them, and stuffed them into a ballot box while incredulous election officials stood by. Later, unable to separate the women's votes from the men's, the officials included them in the totals. The women were later arrested and jailed for election tampering. President Jonathan Allen apparently supported his wife's actions completely, though another husband, anecdotally at least, left his wife in jail overnight. Tomorrow C. B. would follow Abigail Allen's path with no husband to bail her out.

Meanwhile, Pam wondered whether tomorrow's events would turn violent. She remembered her 1965 march in Montgomery. Along with a few friends from Chatham College near Pittsburgh, she boarded a bus and went south after seeing television reports of police with fire hoses, clubs, and dogs attacking black demonstrators in Selma. When she joined people marching to downtown Montgomery, they were routed by club-swinging whites on horses and by motorcycles roaring through the crowds. At one point, while sitting in the road with other civil rights protesters, she suddenly understood the point of the demonstrations. White college kids were supposed to be beaten alongside the blacks in order to generate nationwide sympathy. In the context of the South in the 1960s, integrated protests were enough to ensure violent reactions.

C. B. looked at Pam and laughed, "Let's get into our hay loft and get to sleep."

People began arriving at the Miller house at dawn. Stuart Campbell had them park their cars behind the house, so the technical team would not suspect that protesters were waiting for them. He thought that if they came to Allen, they would follow the procedure the siting commission had outlined in the local papers—knock on the door and ask Don Miller to accompany them onto his land.

At the gathering around the kitchen table, Stuart explained the plan. "If we stay hidden, we will enjoy the element of surprise and can ambush them when they come. When we see the technical team pull up to the house, slip out the back door and get into your cars. Don will answer the door and invite them into the house to discuss their plans. When he does that, we can ambush them. I'll signal you to pull your cars around front to hem them in."

"When do you want me to call the people at the other sites and let them know that they're here?" asked a woman who lived near the site.

"As soon as they park in front of the house," Stuart replied. "But tell them to send only one-half the troops until we're sure that they're all here."

"What should we do after we block their cars?" asked Tom Green, another local ACNAG member.

"Try to slow them down any way we can," Stuart responded. "It won't take long for the people to get here from the other sites. Talk to them. Ask them to show their credentials. We need to hassle them until enough people get here so that we can surround them and keep them from conducting their walkover."

"Maybe we can make a lot of noise out back while they're inside talking with the Millers," suggested Carla Coch, who taught English with C. B. at Alfred University. "They might not know how many of us are here and hesitate."

"If we can believe the newspaper stories," added George Hrycun, another protester, "the sheriff will be on his own with the siting commission. They might think twice before trying to get on the Millers' land, especially after Don tells them they're not welcome."

"I don't think that we should count on them stopping here," Pam argued. "And what makes you so sure the state troopers won't be around?"

No one was more skeptical about the declared intentions of the siting commission than Stuart, though this time he felt sure that they would follow their announced plan. "They won't try to sneak on the land; they'd lose too much face."

Stuart opened a box that was lying on the kitchen table and held up a pair of handcuffs. "I brought several pair of handcuffs with me. I don't have enough for everyone, but some of you can have them. If the technical team tries to get on the land, cuff one of them to you. If we have to chase them through the woods, use the handcuffs to stop them from doing their survey. That should allow enough time for people to get here from the other sites."

No one else in the room had thought about using handcuffs. While it seemed reasonable from a tactical point of view, the mood became somber as everyone realized the seriousness of their commitment. Apprehensively, people glanced out the window, almost expecting the siting commission to materialize suddenly, but nothing stirred as a bright sun rose over the frozen land.

Carla Coch was the first to take a pair of handcuffs. The somber mood lightened instantly when she asked, "What do you and Sally do with these handcuffs anyway?"

C. B. immediately joined the banter. "You seem to be awfully quick to want a pair, Carla."

Stuart extended handcuffs to C. B. "You'd better take a pair too."

"Not me. Do you know how difficult it is for an older woman like me, who has to pee frequently, to go in the middle of the woods? I'm certainly not going to handicap myself further by being shackled to some man!"

The sheriff also arrived at his office early in the morning. The technical team from the siting commission was going to do its first walkover of a proposed site this morning. Surprisingly, Larry Scholes felt calm and detached. After the meeting with the state troopers six days before when they told him they wouldn't assist him in getting the technical team on site, he had been edgy, focusing on routine matters with difficulty. He had kept replaying possible scenarios in his head. What would he do if an angry crowd surrounded the technical team's car? What would he do if the members of the siting commission became belligerent? What would he do if a landowner met them with a shotgun?

Curiously, now he could concentrate on writing routine reports on the very day that he would escort the technical team to one of the sites. Psychologists would probably say that he was in denial; he felt as if he were in the hands of God and might as well do whatever had to be done. The sheriff wasn't a particularly religious man, but he had been praying a lot recently.

Larry heard undersheriff Bill Timberlake entering his office. He rose from his desk and followed him in. "Are you ready for the big day?"

"I guess so. What do you want me to do?"

"I won't know until we find out what's going to happen. We can talk about the specifics when we get in the car together. I really need to finish this report. You greet the technical team and discuss arrangements with them. See if you can find out which site they're going to and let me know immediately."

As the sheriff was leaving the undersheriff's office, Bill quipped, "I'm sure I'll enjoy the companionship." He knew why Larry wanted the information about the technical team's destination. The sheriff was going to position his two deputies who did civil work in the department somewhere near the scene, but out of sight.

Shortly after the sheriff's department settled into an everyday routine, the technical team arrived. Bruce Goodale, the siting commission's environmental director, headed the seven-person group. Protesters would later describe him as verbally combative and argumentative, succumbing to sarcasm when things didn't go his way. He was accompanied by Kathleen McMullin, the liaison between the commission and the county, whom dump opponents had nicknamed the "ice princess" because she seemed so coolly efficient. The nickname seemed especially appropriate after she claimed that she would be happy to have her children live next to the nuclear disposal facility. The team also included another representative of the siting commission and four employees from Roy Weston Associates, a scientific consulting firm that had been hired to do the technical work: two geologists, an archeologist, and a security officer.

When Bill Timberlake met them at 9:30 A.M., Goodale told him that they would be going to breakfast and returning around eleven o'clock. The security officer asked the undersheriff whether he could supply them with bulletproof vests. Bill replied that he thought he could rustle some up. They told him that they were going to the West Almond site.

Bill informed the sheriff.

County historian Craig Braack had arrived at the Caneadea site shortly before 7 A.M. with his "Green Thumb" worker, Delores Fleming. They drank hot chocolate, joined in speculation about the arrival time of the siting commission, and wondered what would happen. Craig had decided several months earlier that he would remain neutral in the struggle so that he could more objectively record the events that might, he felt, dramatically change the character of the county. Taking slides of the early morning activities, Craig was reminded of pep rallies that he had attended in high school, with bonfires, rousing speeches, and conviviality among folks from all economic and social groups.

At about ten o'clock, Craig had misgivings about remaining at the Caneadea site. He turned to Delores and said, "Let's take a chance on this crapshoot and go back to Belmont." They hopped in the car and took the most direct route back. Craig knew that Larry would be escorting the technical team onto the site and hoped to catch the technical team before it left the sheriff's office. He could then follow them to one of the sites.

Craig and Delores arrived at the sheriff's office shortly before 11 A.M. and were surprised to see Larry and Bill working in their offices, acting as though this were just another routine day in the sheriff's department. The sheriff's office was in a corner of the ground floor of the county building. Craig and Delores sat on a bench in a lobby area.

At eleven o'clock two rental cars, containing the seven-person technical team, pulled up to the building. As the team came into the building, Craig rose and quickly introduced himself to a couple of its members. "My name is Craig Braack and I'm the county historian. My assistant, Delores Fleming, and I will be taking slides for the county's historic record. I will be careful not to get in your way and have no intention of impeding your progress."

"Have you been out to any of the sites this morning?" Bruce Goodale asked casually.

"I was up at the Caneadea site."

"How many protesters were up there?"

"When I left there were forty or fifty at the Old German Church, but I heard that there were three other checkpoints with people. So by now there could be as many as three hundred people on the site."

As Goodale walked toward the sheriff's department, the county historian saw him turn to someone else on the technical team and say, "Let's go to Allen. We have the court's permission to go to Allen."

The undersheriff watched the technical team enter the office and rose to greet them. "I have the bulletproof vests that you asked for over here." As they were removing their outer clothing and donning the vests, the sheriff emerged from his office and joined them, his coat slung over his arm.

A geologist, struggling to put on his vest, turned to the two law enforcement officers and asked, "Aren't you going to wear vests?"

Bill answered, "No."

Someone else asked, "Why not?"

Bill answered, "Why should we? The protesters aren't mad at us."

Members of the team glanced at one another. The geologist muttered, "Oh, wonderful."

When everyone was ready, the sheriff started to leave. Goodale then told him, "We've been talking it over and we're going to go to the Allen site."

"That's fine with me," the sheriff responded. "I don't care where we go. I'll stop when we get to the intersection at Aristotle which is next to the site and you can tell me where to go from there. One more thing—Bill and I will do everything we can to get you onto the site, but you must follow my orders. If I perceive things are getting too dangerous and suggest that you leave, you must do just that. You know that the state police won't be around."

Before leaving the department, the sheriff stopped at the office desk and told the secretary to inform his two civil deputies, the sheriff's only backup, that they would be going to Allen instead of West Almond.

Bill Timberlake, walking a few paces behind the sheriff, was surprised that the technical team had changed its destination. Bill concluded that the siting commission had changed its plan, because they didn't trust the sheriff's department to keep their goal a secret. He smiled to himself when he saw Alton Sylor, the county legislator from Allen, in the lobby of the building.

A retired dairy farmer, Alton Sylor was probably the most fiscally conservative legislator in the county. A small wiry man with a face chiseled by the elements, he always opposed spending taxpayers' money unless it was absolutely necessary. Nevertheless, he had kept uncharacteristically quiet when some of his colleagues talked about following the example of Cortland County, where the other two finalist sites were located. The legislature there had created an office and hired a full-time specialist to coordinate their county's fight against the nuclear dump. Although Sylor was angry enough at the state's intrusion into the county to appreciate the tough stance of the protesters, he was too conservative to fully condone breaking the law through civil disobedience.

When he saw the legislator, the undersheriff fell back a few more paces, slowed down, and nodded. "Well, we're off."

"So, are you going to West Almond?" Sylor asked.

"No, we're going to Allen."

"Allen? Oh . . . Okay."

Timberlake smiled to himself. "The siting commission was right to distrust the sheriff's department a little," he thought. There would be few secrets in Allegany County.

The two officers got into their car, and Bill drove to the parking lot's exit. The technical team pulled their two rental cars up behind. The sheriff noticed a press car behind them, followed by the county historian in the rear. This would be quite a parade.

"Drive slow, Bill, we don't want anyone to have an accident on these icy roads." That was fine with the undersheriff. Maybe he would drive slowly enough for the protesters to get their troops to Allen by the time they arrived.

Although Larry and Bill had talked extensively about possible scenarios during the previous week, they hadn't discussed much today. Larry reminded

Bill that he wanted him to stay with the car and remain in radio contact in the unlikely event that the technical team got onto the site.

"I hope that the citizens noticed the interview I gave to the newspapers a couple of days ago. If they block the way, as I expect, I'll have to explain to the siting commission and the press that the state police in Albany are unwilling to get involved and that we don't have any extra people in the sheriff's department to handle the job of getting the technical team onto the site. We'll then back off and go back to Belmont. Then it'll be their move to figure out how to get the siting commission onto the land."

In the newspaper interview Larry had expressed anger both at the state police superintendent, Thomas Constantine, and at the governor for ordering the state police "to take no direct involvement" in escorting the siting commission. "That leaves me," he observed, "in an awkward position; it leaves us out there alone. The only people I can spare are me and the undersheriff." He explained to the reporter that if even a few citizens blocked them, as promised, two police officers "obviously won't be able to escort them onto the site." Then, "I will be advising the siting commission staff that they should explore other civil remedies."

He ended the interview with a mild warning to the siting commission: "If the . . . technical staff uses good judgment, they will not put themselves in a position of danger." He also had a message for the protesters: "I also hope that the citizens' group will not put themselves in a position where they have to be arrested if they can simply turn them away from the sites." Unfortunately for the sheriff, things would not go as he hoped.

Bill Timberlake drove slowly up County Road 15. Driving on most Allegany County roads, you wend your way along ancient river beds between hills that rise up a couple of hundred feet on both sides. People in Allegany County are for the most part valley dwellers. Geologically, most of the county is part of a vast plateau which rose a little over one thousand feet above sea level sometime before the last Ice Age. Rivers and streams running in every direction had carved narrow valleys into this ancient plateau. Then glaciers pushed southward into the valleys, deepening and widening them, leaving glacial till in terminal moraines throughout the area. In most places, if you climb up the hills that rise above the valley roads, you can see for miles in every direction. The tops of the hills are all about the same height.

Around noon, the sheriff and his deputy finally reached a junction in the road, known locally as "Aristotle," with its historic Stage Coach Inn, built in 1817. Now the family home of Dorothy Chaffee, it had been an important stop along the salt mine trail from Livingston County to the Allegany River in western New York. The two officers were surprised that they had seen no sign of protesters along the route. Bill pulled the car onto the side of the empty road, leaving plenty of room for the trailing vehicles to pull up behind him. The officers waited in their car as members of the siting commission got out to look at maps. The sheriff saw somebody point toward the Chaffee house across the road and Bruce Goodale went over and knocked on the door. Bill and Larry got out of the car and followed.

Dorothy Chaffee stepped out onto her porch. She was the matriarch of a family with four generations currently living on or near the site. Three generations had lived there before her. Her grandson who lived down the road was operating a dairy farm on some of the land that was part of the proposed dump site.

Bruce Goodale asked her something and she pointed down the road. Just then she recognized Larry Scholes and nodded hello. She turned back to Goodale, wagged her finger at him, and scolded: "I'm eighty-two years old and this is the most unfair thing I've ever seen in my life. It seems like you can take my home and I haven't any say in the matter. What you're doing to us is worse than what's going on in eastern Europe."

Goodale apologized for disturbing her, mumbled his thanks and backed off the porch. Dorothy Chaffee followed him down the steps and continued to badger him. "I guess I'll just become a bag lady, because at my age it's really difficult to uproot and move. I've lived here for fifty-eight years." She paused, watching Goodale retreat, turned to go back inside her house, slipped on some ice, and fell to the ground. Larry and Bill, a sick feeling in their stomachs, rushed up to aid her. Slowly getting back to her feet, she assured them that she was okay. They helped her into the house and stayed a couple of minutes to confirm that she was all right. The sheriff hoped that nothing more serious would happen today.

Craig Braack, the county historian, was standing outside his car when Goodale went up to the Chaffee residence. There had been no cars on the road since their arrival. Now, a lone vehicle approached. Braack recognized the driver and surreptitiously motioned for him to slow down. Not wanting to compromise his stance of neutrality, he hoped the siting commission was not looking. In a hurried and hushed voice, he told the fellow to let the demonstrators know that the siting commission was at Allen, then quickly waved the car on. The driver, however, was not part of ACNAG and didn't know that protesters were waiting only a mile down the road at Don and Gloria Miller's house.

Everyone got back into their cars, the sheriff still not sure where the siting commission was going. The technical team drove about one-half mile past the Millers' house and parked alongside the road next to a snowy field that sloped gently up a hill. A snow-covered dirt track ran across the field at an angle. The property was owned by William Giovanniello, whose house was a hundred yards away. Leaving his undersheriff in the car, Larry Scholes walked up to the technicians, who informed him that they would be going onto the site at this point.

The sheriff addressed all the members of the technical team. "There are a few things we need to get straight. What you are doing on the site is your business. My business is to keep everyone safe. So if I tell you to do something, please do it. I will only tell you to do something if I judge it to be important for your safety. If, for any reason, you do not follow my directives, you won't have my support in the future. Finally, you'll be going into the woods and I do not want anyone to get separated and lost. So please stay together."

They nodded their heads and looked out into the desolate landscape. Since leaving Mrs. Chaffee, they had seen no other humans. No cars had passed. No demonstrators had greeted them. The eerie silence and the sheriff's warnings slightly unnerved everyone. After comparing their maps to the surrounding territory one last time, Bruce Goodale led the group into the field at around 12:30 P.M. They left their security officer behind, standing beside the cars.

Bill Timberlake remained in the sheriff's car, maintaining radio contact with Larry according to plan. He couldn't believe that no one had shown up to confront the technical team. Impulsively, he reached for the radio and announced to whomever was listening out in scanner land, "The siting commission has just entered Bill Giovanniello's property near his house."

6 Linked Arms

Allegany, rise to your feet.
Lock arms and stand for the people and the land you love.
Never let the poison spread.
Those who lie would see you dead.
No radiation without representation.
Let the ones making it,
know people just aren't taking it.
We're not expendable and neither is the land.

—B.A.N.D.I.T.S. song by Sue Beckhorn,
Co-Chair Concerned Citizens of Allegany County

DECEMBER 13, 1989, AFTERNOON—

PEOPLE ON ALL THREE SITES became increasingly anxious as the day wore on. A little before noon, fearful that the technical team had somehow slipped onto the Allen site, Stuart suggested to Bill Coch that he go out to check the site. The doctor had come to the county as part of the National Health Service Corps, a federal government plan to entice doctors into rural areas where there were not enough physicians. A portion of the doctors' medical school debt was forgiven each year that they remained in the corps, service that would also fulfill any future military obligation. Bill and Carla, his wife, had chosen to go to the village of Andover.

Although rurally isolated, the village was only ten miles from Alfred University. They hoped that the intellectual and cultural stimulation of the university would make rural life tolerable. Both fell in love with the area and the people, and they became active in their communities. Carla not only taught part-time in the English department at Alfred University but was currently serving as president of the Alfred-Almond School Board. Bill had a large private practice, served as the medical consultant to the County Health Department,

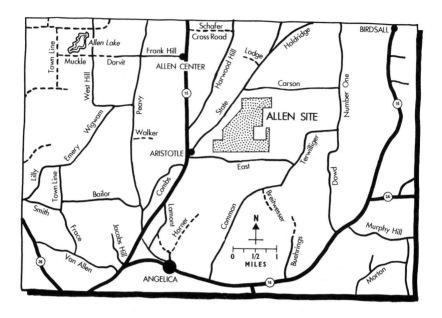

Allen Site Map design by Craig Prophet

and had just been named the medical director of Jones Memorial Hospital in Wellsville.

As Bill started to leave the Millers' house, Stuart suggested that Marilyn Ostrander accompany the doctor to check out the site. Since Marilyn lived there, she would know where the siting commission might have pulled onto some other farmer's land and hidden their vehicle. They drove toward the Chaffee house and began to circle around the area, stopping briefly to exchange notes with Drew, who was parked on the high side of the site next to state forest land.

They had nearly completed the loop when they noticed cars parked along the road. Bill looked across a field and saw people disappearing into the woods. He slammed his car to a stop and handed the keys to Marilyn. "Go back and tell people they're here. I'm going to go up in the woods and chase them." He jumped out of the car and tore across the field.

Upon entering the woods, Goodale pulled out his map, trying to figure out where they were. The county historian, who had followed the group up into the woods, watched as somebody picked up a stick and poked it through the snow. Another kicked at a stone. They were as much out of their urban element, Craig Braack thought, as were the Spanish explorers who had been searching for the Fountain of Youth in the swamps of Florida. The sheriff could see no pattern to what they were doing, and he wondered whom they were trying to impress. They certainly weren't persuading him that they were engaged in some important study. He supposed that they were trying to create a convincing show for the two newspaper reporters who had tagged along.

The people on the technical team were bending over a map when Bill Coch burst onto the scene. "Mr. Miller sent me up here and said that you are to leave his property right now."

Bruce Goodale nervously glanced up at Bill and responded. "The courts have given us the legal authority to do our studies on the site. We don't need the owner's permission."

"But you don't even know where you are. Show me the border of the property."

Bruce turned his back on the doctor and tried to ignore him. Bill wondered how he could singlehandedly stop the technical team from exploring the property. Ironically, at the last ACNAG strategy session, Bill had tried to get others to consider seriously how one or two protesters could nonviolently hinder the siting commission should they get onto the land. Everyone had laughed and dismissed the question, saying it was an extremely unlikely scenario. Now he could think of nothing better than to harass them verbally.

"Why don't you have to get the owners' permission? For that matter, why don't you have to get the signed consent of everyone in this area about their willingness to accept the increased risk of cancer that'll come from the allowable twenty-five millirem release of radiation?"

Bruce turned back to the doctor. "The risk is minuscule."

"Whose idea of minuscule? Yours or mine?"

"It would be a minuscule amount," reiterated the leader of the technical team.

"The risk of childhood vaccinations is even more minuscule, yet I have to get permission to immunize someone. Why don't you have to get consent to be here?" asked the doctor.

"The landowners were notified that we would be coming out here," Bruce replied.

"But did you have their consent?"

"We notified them." Goodale started to turn away again.

"Have you read the BEIR Five report that came out in the last two weeks? The allowable limits of radiation exposure that you are taking into account were based on BEIR Three. It appears that the health risks for low dosages of radiation are three to four times greater than stated in the previous reports." Bill was referring to a report that had just been released by the National Research Council, an arm of the National Academy of Sciences. The

BEIR (Biological Effects of Ionizing Radiations) Report concluded that the "cancer estimates . . . are about three times larger for solid cancers (of the breast, lung, stomach, ovaries and other internal organs) and about four times larger for leukemia than the risk estimates presented" in the earlier report.

"We can easily make the adjustments necessary to account for the new report."

"But have you read it?"

"Not yet."

The doctor continued badgering the technician. "Don't you think that you should read the report, before you say that you can easily adjust your specifications? And why should I have to get consent for exposing someone to a minimal hazard and you don't have to? Shouldn't you get the consent of people in this area, before you are allowed to put a nuclear dump here?"

As the exchange got more heated, the sheriff wondered if he would have to break up a fight in the woods between a doctor and a scientist. He knew that Bill Coch was fairly cool, but Bruce Goodale was becoming agitated. Scholes took a couple of steps toward the two men. Bill noticed the sheriff watching him and lowered his voice. He also remembered the handcuffs in his pocket and thought about handcuffing himself to someone. But to whom? He certainly wasn't going to cuff himself to Bruce Goodale or the two women. He needed to find someone who didn't look too threatening and decided the photographer would do. The doctor had mistaken the county historian, who was taking photographs, for a member of the technical team. Two things made him hesitate. First, he knew that handcuffs would greatly complicate the sheriff's job. Second, Stuart had kept all of the keys and was nowhere in sight.

As the doctor considered using the handcuffs, a couple of people, hidden nearby in the woods, began howling like wolves and hooting like owls. Occasionally the noises sounded like Indian war whoops. Other people had found the technical team and were beginning to harass them. Then Coch heard a gunshot in the distance. Don Miller had walked to the edge of his property and fired a shot into the ground.

The sheriff was simultaneously getting information from his undersheriff on the walkie-talkie. "Cars are coming up behind me, Larry. They're parking alongside the road and some people are going up the hill to follow your tracks."

The sheriff told the technical team that it was time to go. "Pay attention and don't get separated from me. Stay with me at all times." He told his undersheriff that they would be coming down. Bill Timberlake replied, "Sheriff, it's getting a little warm down here." The shivering sheriff joked, "Thank God it's warm somewhere." As he began walking down the hill, he felt the hand of a very frightened member of the technical team, probably the young female archeologist, grasping his green parka and holding on for dear life. She didn't let go until they reached the road.

Alton Sylor, the elderly conservative county legislator, burst onto the scene at the Miller house. "What are you guys sitting here waiting for? The siting commission is just up the road." Stuart Campbell was nervously pacing around behind the house, waiting for Bill Coch and Marilyn Ostrander to return.

Stuart turned to the others. "Call the other sites, Liz. Tell them to send half their people here until we find out whether all of the technical team has come to the Allen site. The rest of you get in your cars. We'll block their vehicles as we planned. After we get there, we'll decide whether to wait for reinforcements or go after them."

The folks at the Miller house drove madly up the otherwise deserted road. Marilyn Ostrander, driving the doctor's car back to the Miller house, was surprised to see her brand-new car zipping past her, driven by Carla Coch. They haphazardly pulled their cars around the deserted vehicles. Someone in a big station wagon misjudged the distances and slammed into Pam Lakin's new Chevy Nova and tore off the front fender. The first casualty, she thought, staring dumfoundedly at John Reese, one of the locals at the Miller house who was getting out of his old clunker.

Psychologists of stress hypothesize that people living in the modern world physiologically react to danger much like our ancestors who lived in caves. In the old days adrenalin rushed into the system, preparing the body to fight or flee from an attacking tiger or wild boar. Since modern dangers are usually not so physical or direct, we stifle these impulses and thereby induce stress. Whether this hypothesis is accurate or not, it is certainly dangerous to drive a car when one's body signals fight or flee.

John Reese had serious medical conditions, including a rare form of cancer, that his doctor believed were caused by exposure to radiation. While working as a janitor in the animal laboratories at the University of Rochester Medical Center, he had been exposed to at least one radiation spill in a laboratory run by David Maillie. Among the experiments conducted in that laboratory was one that overexposed dozens of beagle puppies to radiation in order to analyze the cause of their death. Ironically, Maillie was now one of the five siting commission members in the state of New York who had determined that the land across from Reese's retirement property was a prime site for a nuclear dump.

Meanwhile word had spread to the Caneadea and West Almond sites that the technical team was at Allen. It was only with difficulty that Spike Jones was able to keep a skeleton crew on his own site as people jumped into their cars and set off at breakneck speeds toward Allen. Discipline was a little better at West Almond, only because half the people had earlier responded to a false alarm, rushing off to Caneadea. They wouldn't make the same mistake twice.

After sending half his people to Allen, Gary Lloyd at West Almond kept the rest lined up in their cars until the news was confirmed. While waiting, a fellow whom Gary had never seen before drove up next to him in an old pickup truck and yelled out the window, "I heard what's going on. The state people are over at Allen, huh?"

Gary responded, "That's what we're waiting to find out for sure."

"Well," the man said, "I want you to know something. I'm ready to go any time you guys are. I brought everything I've got. Here it is in the pickup. Maybe somebody needs something."

Looking into the truck, Gary saw three rifles, two shotguns, and a lot of ammunition. He was momentarily stunned.

"It's all I've got," the fellow reiterated, almost apologetically.

"Listen," Gary said, "we're not into guns. We're going to stop them with civil disobedience and nonviolent protest. We don't want any firearms— absolutely no firearms anywhere near the civil disobedience action."

"Well, I don't know if I'm into protesting. I thought you guys were going to fight the siting commission and keep them off our land. I don't know anything about civil disobedience."

"We think we can stop them from doing their work through nonviolence," Gary responded. "If you want to come with us, leave your truck and your weapons right here and get into my car." But the man put his truck in gear and drove away. Gary never saw him again.

Approximately fifteen people gathered around Stuart Campbell near the encircled vehicles. He addressed them, "We've got their cars. They're going to get cold; they'll want to leave and we can take them when they come out. Or we can go in after them." Only Stuart was ambivalent. Without exception, everyone else was eager to track them down. Stuart saw that he couldn't restrain them long and simply said, "Make Allegany County proud. Don't touch, shove, or handle anyone. When you find them, make a circle. I'll get there as quickly as I can. Remain silent; I'll do the talking."

Walt and Peter Franklin, the earliest arrivals from the Caneadea site, immediately took off, wading through the snow-covered field. The two brothers had spent much of their childhood in upstate New York. Both had graduated from Alfred University. Walt was an essayist and poet, Peter a novelist who had taken over the small family farm. Both were serious environmentalists. Walt had spent four years teaching school in Virginia, but chose to return to the area with his wife Leighanne. They bought a piece of property near the village of Greenwood, located in the southeastern corner of the county where rugged hills began to form the Appalachian Mountains that extended into Pennsylvania. A couple of years earlier, Walt undertook a hiking, fishing, writing project. For a year he fly fished in the area, writing a poem about each stream that he explored. He eventually published them in a book.

A group of fifteen people followed the two brothers across the terrain. Stubbles of corn stalks and scraggly remnants of goldenrod jutted from beneath the snow, giving the field the appearance of a man with a three-day growth of beard. The snow had settled and people could run only by lifting their legs high and prancing like deer. Traversing the woods was more difficult; snow drifts were two feet high in some areas. In other places snow had crusted across creeks and fallen logs to waist level. The protesters, like hounds after a fox, fanned out, stumbling and rushing to catch the technical team. After ten or fifteen minutes of tramping through the brush, they paused, looking for tracks in the snow, only to hear a roar from people down the hill at the road. Slipping off logs and falling into snow banks, the trackers tumbled pell-mell toward the sound, flushing two startled deer ahead of them.

At about 1:30 P.M. the technical team, led by the sheriff, spilled out onto the road about one-quarter mile south of their vehicles. Forty protesters met them as they headed toward their cars. Larry raised his arms to part the crowd, but the people linked arms, closed in, and surrounded most of the members of the technical team. As the trackers emerged from the woods, they saw a geologist gawking at the scene in front of him. He reminded them of a deer frozen by headlights. They immediately surrounded him.

"What are you doing?" asked the startled sheriff.

"No one can leave this circle," someone shouted.

"Are you telling me that I can't leave?" Larry Scholes asked.

"That's right. No one can leave or enter this area," exclaimed a woman. Someone else contradicted her statement. People began shouting various directives.

"I would like to talk with one spokesperson," Larry said very calmly with all the authority he could muster. He pulled out his walkie-talkie and called his undersheriff. "Bill, the citizens have surrounded the siting commission. I think I'm being held, too."

Stuart Campbell, a middle distance jogger, had run the quarter mile from the technical team's cars just in time to hear Larry talking with his deputy. "That's not true, Larry. You're free to come and go. We're only holding the siting commission."

"The technical team hasn't violated the penal law," the sheriff calmly stated. "They have a court order to go on this land. You've made your point loud and clear. Civil remedies should be explored further."

"I understand your point of view, but we're going to keep them under citizens' arrest since they penetrated the site," Campbell replied. "They've trod on sacred ground. We had hoped to keep them off the land and would've preferred to avoid arrest."

"Are you prepared to be arrested for this?" asked the sheriff.

"Yes we are," Campbell responded.

The sheriff relayed the information to Bill, telling him to do nothing for the moment. "The citizens say they're going to keep the technical team in the circle, but I'm free to move about." The chaos of the first few moments had slightly unnerved him. He had confidence that the protesters would not react rashly, but he also knew that people often acted out of character in a crowd, especially when confusion existed.

Protesters continued to stream into the area and the circle expanded, providing a larger buffer zone between the technical team and the demonstrators. Larry turned to Kathleen McMullen and Bruce Goodale who were looking at him. "It's very important that you all remain calm. You're not in any immediate danger and shouldn't do anything to provoke the situation." Noticing that television crews from Buffalo and Rochester had arrived and were recording his words, he added, "I know that Governor Cuomo and Superintendent Constantine have decided that the state police shouldn't get involved, but I'm sure that they'll do their duty when they learn of the situation. I'll be contacting them shortly." He had already decided that his only leverage with the state police was to embarrass them in the press, and he wasted no time doing just that.

Surveying the scene, the sheriff noticed that another group had surrounded someone just beyond the large circle. It would make more sense to merge the circles so that he could deal with a single situation. The protesters cooperated. The two cells merged, the geologist joining his colleagues in the center of an expanded circle. Larry took a deep breath and relaxed when order was established. He took out his radiophone and called the state police.

Lieutenant Charles McCole and Captain Robert Browning arrived at 2:00 P.M. Before they could begin to talk things over with the sheriff, microphones were shoved into their faces, while television crews recorded their comments. "When are the state police going to begin the arrests?" asked one reporter.

"There's not necessarily anything criminal taking place here, so there's no reason to make any rash moves," responded Lieutenant McCole.

Another reporter asked, "How many troopers are in the area? How many will be coming?"

"I don't know," replied McCole. "I might be enough. Everything looks nice and peaceful here."

The sheriff believed that Lieutenant McCole and Captain Browning had probably jeopardized their careers by responding to his call for help. He knew he had placed them in a real predicament. The sheriff couldn't imagine their refusing a plea for help, yet he also knew that, short of violence, they were under strict orders to let the sheriff handle things alone.

McCole approached the ACNAG leadership in the center of a circle that now included nearly one hundred people. Stuart, Spike, and Gary, the three site coordinators, would be negotiating with the police. Sally Campbell, media spokesperson, and I joined them. I had been coordinating the nonviolent training sessions throughout the county and had become the moderator for the large ACNAG strategy sessions. I spent the rest of the day filtering information from the negotiating team to the rest of the protesters and making sure that everyone in the circle understood our commitment to nonviolence.

"What is it that you folks want?" McCole asked.

The ACNAG leaders looked at each other dumfoundedly. They had failed to keep the technical team off the land and had only caught them after they emerged from the woods, heading back to their cars. McCole saw that his question caught the leaders off guard. He, too, appeared bewildered and a bit unsettled; he apparently would have preferred reacting to a clear demand from the demonstrators. The protesters were surprised that the state trooper seemed to be negotiating with them. By keeping the technical team from going about its business, they expected to be arrested. "Maybe you'd better figure out what you're doing and what you want," McCole said as he moved over to huddle with the sheriff.

It didn't take the leaders long to solidify their position. Stuart Campbell moved toward the three law enforcement officers and said, "We'll let the siting commission leave if they promise to drive straight out of the county. Otherwise we'll hold them until dark when they will no longer be able to conduct their business." Now the officers seemed bewildered. Stuart Campbell wondered why they seemed so hesitant and blurted out the obvious, "Of course, we're prepared to be arrested."

The three officers left the circle and tried to find someplace where they could talk without newspaper and television reporters eavesdropping. Apparently realizing that the nearby Miller house was hostile territory, Captain Browning told McCole and Scholes that he would drive down to the Chaffee residence and call police headquarters. Only fifteen minutes had passed since the state troopers had arrived. The other two officers soon realized that they would be pestered by reporters if they remained in the area. They walked the quarter mile toward the sheriff's car where Bill Timberlake was patiently waiting. Just outside the car they saw a small circle of demonstrators surrounding the security officer from the Weston consulting group.

"Are you all right?" the sheriff asked.

"As well as I can be under the circumstances," responded the man who had stayed behind when the original expedition had gone into the woods.

"Would you like to go join the larger circle?"

"No, I'm fine here. These folks have been friendly enough under the circumstances. And your deputy is just inside the car."

The sheriff thought that everything had settled down and none of the technical team were in any danger. He and the lieutenant decided to continue walking the two miles to the Chaffee residence.

Freddy Fredrickson, a technical specialist in charge of building kilns for the ceramics college at Alfred University and one of the protesters encircling the security officer, told the man that he had definitely made the right decision.

"Why is that?" the man asked.

"Because it's at least twenty degrees colder in the other circle."

The security officer looked puzzled. Freddy paused, then explained, "Kathleen McMullin is up there." Everyone chuckled. The good-natured banter strengthened the human bond between protesters and captive.

Protesters in the main circle were astonished to find that the police presence had entirely vanished by 2:30 P.M.; they had expected the state troopers to return en masse to begin the arrests.

Stuart called out to me, "Tom, I think the arrests will be coming any minute. Make sure everyone's prepared." I went around the large circle to confirm that everyone was ready and understood our commitment to nonviolence. Those who were willing to be arrested were wearing orange surveyor's tape around their arms, while the supporters wore yellow armbands. I reminded everyone that we would call ourselves "Allegany County" when the police asked for our names, forcing the police to transport us twenty-five miles back to Belmont rather than simply issuing appearance tickets on the spot.

In making the rounds I noticed three men wearing red rather than orange armbands. "What do the red armbands mean?" I asked.

"We don't believe in this nonviolence bullshit. Red means we're willing to use violence to stop the bastards."

"All these people," I indicated the others in the circle and the supporters beside the road, "are here because they want to stop the siting commission.

They've all come with the understanding that we'd be stopping them through nonviolent resistance. We can't have any violence here today."

"We'll go by your rules for now, but you won't stop the state this way. Down the road, we'll do sabotage and even pick up guns."

"There isn't time now to discuss the philosophy of nonviolence, but we think it'll work. The state has a hell of a lot more firepower than you do. Nonviolent strategies will unify the county; violence will frighten and divide us."

"It's not going to work," said another fellow, "but we'll go along with you today."

I gratefully noted that the guys with the red armbands were standing next to the husky Franklin brothers. They had heard the whole exchange; Walt had attended the first meeting in Gary Lloyd's basement. I knew that if anyone could relate to the people with the red armbands, they could. Walt looked at me and nodded as if to say that everything would be cool.

I also noticed that one of the team members, a young archeologist from Dunn Geoscience in Albany, was standing nearby and probably heard the conversation. Inadequately clothed, she was shivering. She also seemed very frightened. I alerted Stuart that she might be suffering from exposure.

I was surprised to see Glenna Fredrickson, treasurer of CCAC, in the circle. CCAC's lawyer had warned the officers not to get arrested. Steve Myers, the chairman of CCAC, decided to stay away from the protest altogether, though his wife Betsy was in the circle linking arms with other ACNAG protesters.

Glenna had been raised in Tennessee for the first eighteen years of her life and had recently moved with her husband to Allegany County. She now wanted to be arrested, even though she had taken an oath of office as an elected member of the Alfred Town Board to uphold the law. Hating the cold northern winters, she was wearing a huge down coat, snow boots, and battery-powered socks as the midafternoon temperature was rapidly falling toward zero degrees fahrenheit.

Sue Beckhorn, vice president of CCAC, turned to Glenna, "You can't get arrested, Glenna. You're the treasurer!"

"You're in the circle and wearing an orange armband," Glenna countered.

"Yes, but if you're arrested, who's going to bail us out of jail. You control all the money," replied Sue.

"But damn! I want to do it. This feels right." Glenna paused and reflected. "Okay, but I'm going to stay in the circle until the police return."

An hour passed without any sign of police, except for Bill Timberlake in the sheriff's car a quarter of a mile away. Supporters passed out coffee and hot chocolate to both protesters and members of the technical team. Reporters went back and forth through the linked arms of the protesters as though they were passing through the permeable membrane of a one-celled creature. People stamped their numbed feet on the snow-impacted road, while supporters brought cardboard for people to put underneath their feet to help insulate them from the frozen turf.

Sue Beckhorn, wearing her Revolutionary War tricorn hat, taught everyone a song she had recently composed that would quickly become the anthem of the movement: "Allegany, rise to your feet. Lock arms and stand for the people and the land you love. Never let the poison spread. . . ." Her numb fingers did not allow her to play the guitar that she had brought with her. Never was such a simple song sung so out of tune. People would later joke that the protesters probably crossed the boundary of nonviolence by subjecting the siting commission to such acoustic torture. But never again would the words of the song and the protest activity be in such perfect harmony.

Sally Campbell used the lull in the action to talk with reporters. Although Sally had been dealing with the local press during the last few months, this was her first experience with the regional media, including television news reporters from Buffalo and Rochester. She knew that it was critical to have these two media centers cover the story. If cameras were present, she believed, the police would be hesitant to use extreme tactics, and if the story were well covered, it would be more likely to get additional protesters to turn out for future confrontations. During the past week, Sally had learned a lot about the media. She learned to ask for the assignment editors when she called the TV stations. She quickly mastered the sound bite, realizing how important it was to be succinct and clear.

Today Bruce Goodale had put a spin on the story by claiming that the team had successfully walked over the site and accomplished everything it had wanted to do. Sally, like everyone else, knew that they hadn't accomplished anything. In fact, protesters who followed their footprints in the woods were not certain that the team had even gotten onto the site itself. The siting commission had only spent thirty or forty minutes from the time they began trudging across the field until they stumbled back onto the road.

Sally let them have it. "Spending a half hour on land that was not even part of the site is typical of their scientific procedures"; "Bringing an archaeologist to check for Indian artifacts when there is a foot of snow on the ground illustrates their shoddy work"; "Taxpayers are spending a lot of money to pay for geologists to check out the flow of water when everything is completely frozen." While some reporters broadcast Goodale's comments, Sally's sound bites made them seem ridiculous. ACNAG's media spokesperson began to understand the power of ridicule in her public duel with the siting commission. She also soon discovered that tweaking their noses publicly further solidified the county's opposition to the nuclear dump.

By 3:15 P.M. many protesters became concerned about the deteriorating condition of the archaeologist, who was showing signs of exposure. Shirley Lyon-Bentley, a nurse for a home care agency in Allegany County, got her a blanket and a cup of hot chocolate. She started walking up and down with her in the center of the circle, and she urged the young woman to take shelter in someone's warm car. The archaeologist, however, feared leaving her fellow team members. Even after "neutral" arrangements were made for her to take refuge in the warm van of a television crew, she refused.

The sheriff and Lieutenant McCole walked toward the Chaffee house in silence, lost in thought. Larry wondered whether McCole's superiors in Albany would change their minds and send in the state troopers. Larry had a good working relationship with both McCole and Browning, and he was sure they knew why he wouldn't call mutual aid. He wondered why the state police seemed sure that he would take charge if they did not respond. Now that the protesters were holding the siting commission as prisoners, the state police would certainly have to do something.

The sheriff thought about his choices and still felt he'd made the right decision. He trusted the ACNAG leaders not to do something stupid. The Albany officials were so removed from reality, the sheriff thought, that they did not even consider the financial strain they were imposing on one of the poorest counties in the state—or maybe they didn't care.

The two men entered the Chaffee residence after their forty-minute walk. Browning was talking to someone on the phone. The sheriff assumed that the captain was speaking with his superiors in Albany.

When Browning hung up about ten minutes later, the sheriff could tell that he was not happy with his orders. Browning informed the sheriff that the state police would not become involved beyond his and McCole's presence unless the siting commission was seriously threatened by violence. Larry suspected that Albany was unhappy that Browning and McCole had already involved themselves in the situation. He was astonished, however, that Superintendent Constantine would not order the troopers to intervene after the protesters encircled the siting commission, holding them captive.

The three law enforcement officials got into Browning's car, more uncertain than ever about how to resolve the crisis. The sheriff, still committed to playing chicken with the state police, felt the temperature plummeting below zero. He worried that some of the technical team were not dressed warmly enough to sustain exposure to the Arctic air. The sheriff, unaware that the protesters had resolved to conceal their identities, explored the possibility of arresting the protesters on the spot by issuing appearance tickets. Browning and McCole agreed to cooperate.

The sheriff walked into the circle a little more than one hour after he and McCole had vanished. As soon as Stuart saw the sheriff, he approached him. "You've got to get that woman into a warm car," he said, indicating the young archaeologist. "She isn't dressed well enough for this severe cold. The TV crew in the van over there agreed to let her get out of the cold, but she refused and is acting irrationally."

The woman, a blanket wrapped around her shoulders, was shivering uncontrollably. The sheriff walked over to her. "I can't allow anyone to get hurt here. I must ask you to get into the TV van until this situation is resolved. Please go with this newspaper reporter and get warmed up." The archaeologist complied.

The sheriff walked back to Stuart. "It's getting very cold. You've made your point. Please let the members of the technical team go and pursue civil remedies in the courts."

"You know that we can't do that, Larry, unless the siting commission agrees to suspend its work and drive straight out of the county." Stuart, assuming that state troopers were just around the corner, added, "You'll just have to arrest us."

Before the sheriff could explain his plan about setting up a temporary booking station in the area, Gary Lloyd burst into the circle. "This is Mister Giovanniello, a property owner here, and I think you should listen to what he's got to say!"

The sheriff looked up and recognized William Giovanniello. Born in Avellino, a town near Naples, Italy, he had immigrated to Brooklyn with his family when he was thirteen years old. His mother opened an Italian restaurant and his father worked as a carpenter. Although his parents soon returned to Italy, William stayed in the United States, graduated from high school, and opened a pizzeria in Jackson Heights in New York City.

An older brother had previously started a pizza business in Wellsville. When William and his wife Lorraine visited him, they fell in love with the area. Selling their pizza business in New York, they moved to Allegany County in 1976 and bought a farm. For the next ten years they raised beef cattle while operating Pizza King in the small city of Hornell, just across the Allegany County border.

Although he still owned the farm, his daughter and son-in-law were currently living there. The technical team had not stopped at the farmhouse before entering Giovanniello's property, but the son-in-law saw the scientists traipsing across the land and called the pizzeria, where William was making his special tomato sauce. Still smelling like fresh-baked pizza, he rushed to Allen to find the police planning to arrest the protesters.

"They shouldn't arrest everybody here," he told Gary Lloyd who was standing outside the circle. "How can they arrest everybody here? The siting commission are the ones on *my* land; I'm not on *their* land. So they're trespassing on my land. Arrest *them*."

Gary smiled. "Why don't you go tell the sheriff your thoughts?"

Bursting onto the scene, the naturalized U.S. citizen spoke to the sheriff who was standing next to McCole. "I want them arrested."

At first Larry misunderstood him. "We're trying to work out the logistics of the arrest right now."

"No. No. I want *them* arrested," insisted Giovanniello, pointing to the technical team in the center of the circle. "I didn't get a letter that they were coming. They didn't knock on the door and ask my permission. They were trespassing on my land. Most of the land they were on isn't even part of the site."

"Let me understand this," said the sheriff. "You want me to arrest the siting commission for trespassing on your land?"

Giovanniello nodded.

At that moment a school bus came up the road and stopped at the circle. The driver recognized the sheriff, stuck his head out the window and asked, "Larry, how do I get through?"

In a loud voice, the sheriff said, "We have to step back and let the bus through." The protesters, still linking arms, moved to the side of the road with the technical team still trapped in the middle. When the bus passed, everyone moved back again to the center of the road.

The school bus delay gave the sheriff time to reconsider his options, and he wondered whether there might not be a way out of the stalemate after all. Captain Browning joined the other two law enforcement officers, and Larry asked, "What do you think? If I file Mister Giovanniello's complaint with the district attorney, the citizens will undoubtedly let them go."

"We can't do this," Browning said.

"Well, *I* can do it. I can cite them for trespass," answered the sheriff.

"You've got broad shoulders," shrugged McCole.

"It's much easier for me to do this than it is for you. I'll take full responsibility for it," Larry said, deciding to walk through the door that Giovanniello had opened. "We have to end this situation as quickly as possible."

Turning back toward Stuart, Spike, and Gary, the sheriff asked, "If I were to file Mister Giovanniello's complaint with the district attorney, would you then let the technical team leave?"

The day was wearing on and the sun was just above the horizon. The temperature had fallen below zero. Arctic air numbed the exposed skin on their faces and penetrated into their bones. "If you arrest the siting commission, we will all go home," Stuart Campbell stated, trying to hide his glee.

The sheriff and the two troopers transported complainant and witnesses to the county courthouse in Belmont shortly before 4 P.M. That left the protesters and the technical team once again hostilely staring at one another in the isolated valley. Stamping their feet in an attempt to restore circulation, the jubilant protesters began singing their protest songs—still offkey. In the center of the circle members of the technical team also stamped their feet and glowered angrily at their captors. Kathleen McMullin was walking rapidly back and forth. She paused, and Hope Zaccagni, a transplanted southerner, noticed that she was wearing hush puppies rather than heavy boots. Although the ground rules prohibited talking with the technical team, Hope goaded her. "Kathleen, those shoes are no good for walking through the woods in the winter. Next time you'd better wear better shoes." Kathleen glared frostily back.

Shortly before 5 P.M., a helicopter picked up Wadi Sawabini, a TV news reporter from a CBS affiliated station in Buffalo, and hovered in the air taking pictures of the protest below, before flying the videotape back to the studio. From the air, the scene looked like some archaic rite to celebrate the Winter Solstice.

The sun dropped behind the hills in the west and the sheriff still had not returned. "This is really bizarre," Stuart commented to Spike. "It's now dusk and we could release the siting commission, but the sheriff still isn't back."

"I have never been so damned cold in my life. I should have retired to Oklahoma to raise horses!" Spike saw that Stuart was about to say something, but he cut him off. "And don't give me any shit about this being good for my character."

Stuart, uncharacteristically, decided to let Spike have the last word.

At a quarter past five the sheriff arrived, walked into the circle, and turned to the ACNAG leaders. "Since it was after hours, I called the assistant district attorney at his home. He agreed to accept the complaint. He'll consider the charges against the siting commission. Will you now release the hostages? We've done what we've said we'd do."

Without a word one end of the circle opened, and the exhausted technical team walked toward their vehicles in silence. As a full moon began rising in the east, everyone put their heads back and howled like wolves, "Aaooooow." The county historian shivered at the eerie sound and felt a deep bond of unity surge through the exhilarated protesters. Neither he nor the technical team knew, however, that the impetus for the howl came from a talk that Dave Foreman, founder of Earth First!, had given at Alfred University only a month earlier.

"We'll come in if people are surrounded and held hostage. We'll evaluate other specific situations, but if you request our help, we'll probably come. Browning and McCole, however, may be replaced by troopers from other parts of the state. The superintendent doesn't like your talking about the good local troopers and their bad commanders in Albany. He wants your assurance that you'll stop attacking him in the press and giving the state police a black eye."

"I don't want to attack the state police, but the only power I have is public opinion. I'm not going to let Constantine and Cuomo leave me holding the bag." The sheriff did not want to appear argumentative now, however, since the major had just signalled a shift in the policy of the state police. He lowered his voice and continued, "McCole and Browning have worked with me on many occasions, and we know what to expect from each other in crisis situations. I'd hate to have them replaced. I won't criticize Constantine in the press any more."

"That goes for the governor, too."

"I won't criticize Cuomo either," the sheriff replied.

Bill Coch, Stuart Campbell, Spike Jones, and Jerry Fowler took seats in a conference room in the county courthouse, joining the sheriff, the undersheriff, and three state troopers—Major Kelley, Captain Browning, and Lieutenant McCole. The sheriff had contacted Jerry Fowler, ACNAG's attorney, and told him that the state police would like to meet with the group's leaders. Still insisting that ACNAG didn't have any leaders, the steering committee decided to send Bill, Stuart, and Spike, because the siting commission had already fingered them as ringleaders.[2]

ACNAG and the state police sat on opposite sides of the conference table; the sheriff, undersheriff, and a county legislator sat to the side. After introductions, Major Kelley began the meeting. "I've got another engagement and have to leave in a few minutes. Captain Browning and Lieutenant McCole can speak for me and work out the details. But we can't have another situation

---

2. A few days after the action at Allen, the technical team asked James Euken, attorney general in Allegany County, to charge William Coch, Stuart Campbell, and Richard "Spike" Jones with "Unlawful Imprisonment." On January 9, 1990, Judge Banish, from Allen, recused himself, because his wife Charlotte held an official position in CCAC. When asked by reporters, Euken claimed that although the case was not at the top of his agenda, he was prepared to pursue it. Nothing ever happended to the case, however, and people speculated that Euken did not want to commit political suicide by trying a case against Bill Coch, one of the most popular doctors in Allegany County.

where you take members of the technical team hostage. The sheriff tells me you're all reasonable people and I'm sure we can work something out."

Major Kelley then shook hands and left the room. Captain Browning picked up the conversation. "If you encircle people and restrict their freedom, the state police will no longer stand by, even if you're not physically harming anyone. That's taking hostages, and is considered a serious crime."

Stuart said, "We expected to be arrested and thought the state police would arrive at any minute. Why didn't you arrest us?"

Ignoring the question that had deeply embarrassed the state police, the lieutenant said, "It won't happen again. We're not going to stand by and let you take hostages."

"We weren't trying to take hostages," Stuart said. "We were trying to stop the siting commission from conducting its work."

"We just want to be clear that if you ever encircle people we'll come in to free them," the lieutenant calmly reiterated.

"Encircling them wasn't our plan in the first place," Stuart replied. "That only happened because the siting commission didn't go where they said they'd go and didn't do what they said they'd do. They said they'd ask the landowners to accompany them, but they never even stopped at the Millers' house. They just tried to sneak onto their land."

Everyone in the room shared a common goal. No one wanted to have another dicey situation where anything might happen and people could get hurt. Stuart saw how the meeting might be turned to ACNAG's advantage. "None of us wants things to get out of control. The court ruled that the siting commission has to give landowners three days notice before they show up.[3] If they'd stopped by the Millers' house in the first place, we wouldn't have surrounded them."

The lieutenant stared back at Stuart. "It appears the technical team will have to decide where they're going ahead of time. Does that mean that you'll agree not to take hostages?"

---

3. The court case actually occurred in Cortland County on November 13, approximately one month before ACNAG encircled the technical team at Allen. The judge ruled that the siting commission must notify landowners at least three days in advance—five days if it included a weekend. Although the decision only applied to Cortland County, the siting commission stated in the press that it would treat landowners in Allegany County the same way. Although they had notified the Millers, they never got on their property; instead they went on the Giovanniellos, who had never received any notification. This created the basis for William Giovanniello's filing a complaint against the siting commission. See chapter 5. In February Judge Gorski in Buffalo would formally rule that the siting commission must notify the landowners of Allegany County as well.

"If the siting commission follows the court order and informs the landowners in advance, then we won't surround them. We'll block them from going onto the land, however."

"I'm sure the siting commission will follow the law, so we have an agreement," concluded McCole.

The sheriff had remained silent through most of the discussion, but now reaffirmed his dedication to keeping everything safe and peaceful. "It's extremely dangerous for people to be speeding on icy roads all over the county chasing the siting commission. I'm glad we have the court ruling."

"I want to understand a couple of things," said ACNAG's lawyer. "We all agree that it's unnecessary to hold people captive. But what role will the state police play if ACNAG blocks roads and keeps the technical team from going onto the site?"

"The state police feels that it's the sheriff's job to get the technical team on the site," the captain replied, "but we understand that he has a very small department. If he calls us to help him make arrests, we'll come in."

"That means that you won't use force if the demonstrators don't resist arrest?"

"We won't escalate the situation," said the captain. "We'll do everything in a calm and orderly way. If there's no hostage taking and everyone is cooperative, we won't use force."

Stuart wondered how much control the law enforcement officials had over the siting commission and asked again, "Can we really be sure that the siting commission will obey the court order and go where they say they'll go? Everything depends on that."

"We're here to enforce the law," said the sheriff, "we're not here to take sides. I can assure you that I'm not going to go tearing around the county on icy roads from one place to another. I don't want citizens speeding around the county either. It's better for everyone, including the siting commission, if there are no surprises. So if the technical team wants me to accompany them, they're going to have to proceed in an orderly way."

❖

FRIDAY, JANUARY 12, 1990—

Sally Campbell had spent all day talking to media and had called the Rochester, Buffalo, and Elmira television stations to explain what would be happening. As though fishing for trout, Sally tried to lure camera crews into the county with soundbites: "It's a favorite sport in Allegany County to chase the siting commission"; "We will keep the technical teams off the site if possible—if not we'll keep them from doing any work"; "We'll be nonviolent; the siting commission is the wild card."

She had also spent several hours talking on the phone with local newspaper reporters who were writing major stories about the looming confrontation for their Sunday editions. Assuring the people in Allegany County that the encounter would be nonviolent, Sally emphasized the agreement with the state police. "From our conversations we expect that people will be arrested this time. We expect it to be an orderly process, as it should be for civil disobedience. Everything will be very ritualized. The state police will expect what we're going to do, and we'll expect what they're going to do."

She warned the siting commission through the press that "it's when things happen by surprise that you're liable to have trouble." She emphasized that ACNAG had worked hard to plan for an orderly encounter and hoped the siting commission wouldn't be stupid enough to try to sneak on the land again. "The only wild card is the siting commission," but they're "being forced by public opinion, if nothing else, to play by the rules this time." She reminded the press that members of ACNAG had been disciplined and nonviolent at Allen, even in a very unpredictable situation.

Although masterful in conveying ACNAG's positions to the press, Sally found speaking all day on the phone to be torture. A very private person, she was used to spending days in silence, painting in her studio or remodeling her house. Tonight, exhausted, she fell into bed at nine o'clock and slept more soundly than she had for several days.

Her husband, however, slept fitfully all night long. Stuart awoke a couple of hours after he had gone to sleep, as he had every night during the last week, mulling over the tentative agreement with the state police. He remembered summarizing it two days before at a large ACNAG steering committee meeting. "We agreed not to surround the technical team—the cops said that if we ever did that again, they'd come in and bust heads. They agreed that they wouldn't use force, however, if we were blocking a road; they'd just arrest us in an orderly way. They also agreed not to change plans and dash from one site to another, so we won't have to chase them all over the place; we should know where they're going this time."

The technical team had notified landowners that they would be arriving at Caneadea on the next Tuesday and at West Almond on Thursday. Stuart turned over in bed, worrying the same issues over and over. True, the deal with the police would be beneficial *if* ACNAG could block seven intersections at the Caneadea site, but would enough people be able to take time off from work on Tuesday morning? Seven hundred people had signed up with ACNAG at this point in the struggle, but how many would actually show up wearing the orange armband? Although no one at the steering committee meeting had objected in principle to using vehicles to help block the roads, several wondered whether the police would tow their cars and trucks and impound them indefinitely. How many would make their vehicles part of a blockade? If folks were spread thin, what would keep the technical team from walking around the barricades and directly onto the site? Maybe surrounding the scientists was the only reasonable strategy, after all, but that would lead to violence.

After fretting about these issues for a couple of hours, Stuart, exhausted, drifted back into an uneasy sleep. Maybe he slept for two hours, though it seemed to him like only a few minutes before he awoke again. He had been dreaming about trench warfare during World War I and suspected his subconscious mind was reminding him that he had to start teaching at Alfred University again next week, just when the siting commission would be arriving. He thought about his European history courses and wondered whether he'd have any time to prepare. The bookstore had just informed him that two of the books he was using were now out of print. When would he find time to get to the library to check out substitutes? "This dump business," he groused to himself, "has become a full time job!"

Suddenly he sat straight up in bed. He wasn't teaching World War I this semester; he wouldn't be teaching it until next fall. He looked at his watch and saw that it was 4 A.M. He dashed downstairs, stopping to throw a couple of logs into his Vermont Castings wood stove, and spread a map of Allegany County on the kitchen table. "What if we don't stop them when they're right next to the site, but set up our blockades a mile or so out?" he wondered.

Sally awoke, hearing Stuart rummaging around downstairs. Normally she'd just turn over and go back to sleep, but she began mulling over her conversations with the press and realized she might as well get up.

Seeing her come downstairs, Stuart called out, "Come here, Sally, and look at the map. I've been worried that the technical team would walk around our blockades, especially since they know the police will rescue them. I don't think they'd walk if our roadblocks were far enough from the site. We'd have to go from seven to nine roadblocks, but that would keep them more than a half mile away, about a mile in most places."

Stuart pointed to the map, showing her where he would put the roadblocks. "I woke up dreaming about the advantage the Germans had in World War I, because they controlled the interior and could easily move troops from one area to another. That's another advantage. If we push the roadblocks out, we can shift people more easily from one roadblock to another around the perimeter of the site. What do you think, Sally?"

"If the blockades are right next to the site, I don't see what would stop them from walking. We'd have to surround them and the police would come in and bust heads."

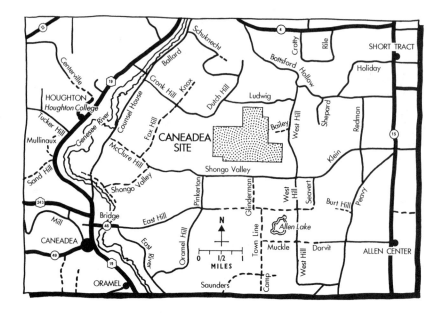

Caneadea Site Map design by Craig Prophet

"Do we have enough troops to set up nine roadblocks? I'm certain that's why the siting commission decided to attack this site first. They were planning to do that last time, but went to Allen, because we had most of our people at Caneadea."

"I don't see that there'd be much difference between setting up seven blockades and nine," Sally responded. "If we can move people around more easily, that'll be a big advantage. Besides, we'll know where they're going to be because Matt Wells and Tracy Moore will be tagging along after the sheriff."

ACNAG had used a few CB radios during the last skirmish with the siting commission, but in the last month they had spent a lot of time and money perfecting their communication system. At West Almond Mick Castle had set up a CB station with a large antenna in the middle of the woods at the top of the site. One weekend, Gary Lloyd's buddies had moved an old chicken coop on a hay wagon to "hilltop," where they put a base radio. Now the station would be able to reach every place around the site and even communicate with the new base at the Eicher's home at Caneadea.

CCAC had spent several thousand dollars to buy sophisticated CB equipment for the new bases and at least twenty individual CB radios. Clearly, CCAC now considered ACNAG its "underground," while ACNAG thought of CCAC as its "front." One of the best ideas to come out of the last ACNAG meeting, Sally felt, was to have someone with a CB radio shadow the sheriff.

"We'll know exactly where they're heading," she told Stuart. "It'll take them at least twenty minutes to back up from one barricade and get to the next."

"That'll give us plenty of time to move people from nearby roadblocks," Stuart said, "and once they begin arresting people, we can easily move more people in."

"I wish Spike were here so we could try these ideas out on him," Sally said.

"I can't imagine he'd object. Anyway, he isn't here to discuss it. I tried to get him to cancel his trip to Michigan when we learned they'd be coming on Tuesday. But he didn't want to, so we've got to plan the details."

Spike had been invited to go to Michigan to speak to people who were resisting a nuclear dump in their community. Protest leaders in Allegany County were not only concerned with protecting their own back yard, but also believed other people should do the same. Leaders of CCAC therefore reached out to other threatened communities, forming alliances with them. Adopting the bumpersticker motto, "Think Globally, but Act Locally," they helped start Don't Waste New York and even tried to found Don't Waste the U. S.

Spike was the inspirational speaker, traveling to many other states where he championed civil disobedience by telling the story of ACNAG's continuing struggles against the siting commission. He warned people to be suspicious of lawyers and national figures who would take control of their local movements, getting everyone mired in lengthy legal battles. ACNAG's lawyer, he told them, was part of the movement and didn't charge anyone to represent them. "Don't spend all your time raising money for lawyers. Certainly don't get involved in discussing technical issues with the nuclear industry! They want to keep you tied up playing their game. Change the rules with civil disobedience. Make them play by your rules. Remember—they want to put a dump in your community and you don't want them to do it. They'll be happy to talk about technical issues until you're too damned tired to fight. Just say 'No.'" Everyone always laughed at that line. Nancy Reagan had coined this slogan to solve the drug crisis.

"Spike's got several people from Caneadea to be in charge of roadblocks, but we'll need to get some more," Stuart told Sally.

During the weekend, Stuart discussed his plan with several people and made sure that all the roadblocks had monitors with CB radios in their cars. He scheduled at least two vehicles to block each intersection. As Tuesday approached, his confidence increased—so did his apprehension. Something was bound to go wrong! Interestingly, he assigned himself to be a monitor at an obscure roadblock where the technical team was most unlikely to enter. Having become deeply involved in the planning, he now happily turned the coordination back to Spike, who returned from Michigan only a few hours before the showdown.

JANUARY 16, 1990 —

Steering her large station wagon into the driveway, Hope Zaccagni noticed a huge antenna rising several yards above the Eicher home at the Caneadea site. "Clearly we're not fooling around this time," she thought to herself. Even though it was only 5 A.M., the lights were already on in the house. She wondered why she had agreed to be in charge of communications at Caneadea. Though flattered that everyone had such confidence in her composure under fire, she had knots in her stomach. Bracing herself against a cold January wind, she opened the car door and walked toward the house.

Entering an immaculately clean kitchen, she saw that everything was already set up. A police scanner rested on a small table in a corner of the room; the CB base radio was at one end of the kitchen table, a large coffee pot with cups at the other. "We're going to be out at a roadblock," Kay Eicher told her. "Neither Ed nor I know much about CB radios anyway. People spent the last few days setting everything up and testing it. They said you could hear as far away as Belmont."

Hope sat down at the table. By six o'clock the roadblocks would all be established and she would test the system to make sure she could communicate with everyone. She read the two-page instruction sheet and began familiarizing herself with the radio. Someone had meticulously organized everything.

Much of the discussion during the last ACNAG meeting had been about radio protocol. No one should use their CB radios for chit-chat. The main band would only be used for communication with the base; side bands should be used for other important communications. People would only follow orders from Hope, who would use her own judgment, or if necessary check with Spike, who was the site commander.

At six o'clock Hope took a deep breath and looked at the sheet of paper with names of blockade leaders and their CB "handles." "Philosopher, can you read me?"

"Affirmative."

She laughed when she saw Stuart Campbell's handle. "Soupy, can you read me?"

"Affirmative."

She quickly worked her way through all nine roadblocks, and then tried to reach the West Almond site. "Hilltop, are you there?"

"Affirmative."

"Are your roadblocks all set up at West Almond?" she asked, even though no one expected the technical team to show up there.[4]

"Yes, we have about six people and a couple of vehicles we can use at each roadblock. I'm glad we can reach each other, because you'll need to send us reinforcements from Caneadea if the siting commission shows up here."

"Okay, and remember that we'll keep the land lines free and can communicate by phone."

Taking a deep breath, Hope finally tried to establish communication with an old red pickup that was parked outside the sheriff's office in Belmont. "Tailgater, can you read me?" Her set crackled, but she couldn't understand anything. "This is where it breaks down," she thought. She repeated, "Come in Tailgater."

"I hear you," crackled the radio faintly.

Hope turned up the volume. "Has the technical team shown up outside the sheriff's office?"

"Not yet."

"Let me know as soon as you see them."

"Roger and out."

"Base to Sarge."

"I hear you," Spike answered.

"You probably heard that I made contact with everyone, including West Almond and Tailgater."

"I heard your end of the conversation. I assume they haven't shown up yet."

"Not yet. I'll let you know when I hear something."

As prearranged, Hope contacted the checkpoints every half-hour. At 7:30 A.M. an excited voice broke the silence. "Tailgater to base! Tailgater to base! The technical team has just pulled into the parking lot in two cars."

"Roger. Base to all units. Tailgater has just informed me that the siting commission is at the sheriff's office in Belmont."

"Well, what d'ya know. They got out of bed early for city folks," Spike said, acknowledging the information. "Looks like they mean business this time."

A half-hour later the caravan set out from Belmont. "Tailgater to base," the radio crackled. "The sheriff's leading the parade, followed by two cars

---

4. Only a couple of people patrolled the Allen site, because the siting commission said they had already accomplished their tasks there—even though they had spent less than an hour on the site.

from the siting commission. The county historian is just behind them and I'm next. There are a couple more cars following us. I think they're reporters."

For the next couple of hours Hope relayed "Tailgater's" information to all units. The technical team approached three roadblocks, only to be turned around. At each one the scenario was nearly identical. The sheriff and Bruce Goodale would both get out of their cars and walk toward the barricade where twenty to thirty protesters waited. Two or three "stalled" vehicles were behind them, the keys hidden in the pockets of ACNAG supporters.

"Who is the spokesperson here?" the sheriff would ask.

"I am."

"The technical team has a court order that allows them to go onto the site. I hope you'll cooperate and let them pass."

"I'm sorry, but we can't do that."

Bruce Goodale would in some fashion reiterate the sheriff's statement. "We have permission from the landowner and would like to go onto the property. We need to do that in order to evaluate whether the site is suitable."

"We understand that. I'm sorry, but we won't let you get onto the land today."

The sheriff would let the exchange continue until he began to feel that Goodale was becoming too angry or the crowd of protesters was growing too large. Then he'd say, "Okay we're leaving. We'll try another road." He saw Sally Campbell talking to reporters at the second roadblock. Worried that there could be a serious accident if people started dashing from one place to another, he went up to her, "Sally, please let the people know that I drive awfully slowly."

The leader of the technical team and the sheriff would then get back into their cars and go to another roadblock, while Tailgater relayed every turn they made. At one point, Tailgater informed Hope, who was following their every twist and turn on a large map of the area, that they had turned left on Peavy Road.

"Left on Peavy Road? What're they doing way out there?" Laughing, she broadcast to all units, "The technical team's lost; it's way out on Peavy Road, going the wrong direction." Weeks later people would jokingly ask the sheriff, "What's wrong, Larry? Don't you even know the roads in your own county?" Peevishly, he'd reply, "Yeah, I know my own county. I turned left, because they told me to turn left. I have to admit they put me on some of the damnedest roads; apparently they couldn't read a map." Not wanting to aid the siting commission, Larry had told them, "Don't tell me where you're going. Tell me to turn left or to turn right."

The caravan traveled on Wigwam and Lilly Roads into Belfast, passing state troopers at a garage in Caneadea, across a narrow bridge that spanned the Genesee River, and on to the fourth roadblock at the top of East Hill Road.

Now Bruce Goodale had tired of the ritual. A little after eleven o'clock, he told the sheriff, "We have to be on the site to do what we want to do. We want to get onto the property today."

The sheriff turned back to Klaus Wuersig, a professor at Alfred State College, who was the monitor at this roadblock. "You are committing an illegal action. We've attempted entrance onto all locations the commission wants. They're adamant about entering the site at this location. If I have to ask for assistance, my assistance is the state police. I don't want to have to do that."

Wuersig replied, "The people you've brought with you serve a criminal and unjust cause. We're ready to be arrested for civil disobedience."

The sheriff returned to his car, while reporters hovered around him. When Lieutenant McCole asked whether anyone was in danger, he replied, "No problem here. We're just blocked off, and the technical team insists on being escorted onto the site."

About twenty minutes later, troopers arrived on the scene. Lieutenant McCole approached the protesters and ascertained that they would not open the road. He informed them that the troopers would begin making arrests.

McCole turned to someone who was blocking the road, "What is your name?"

"My name's Allegany County."

"Mister Allegany County, the technical team has a court order to let them on the site. Will you let them pass and go about their business."

"I'm sorry, but I can't do that."

The lieutenant would then introduce Mr. or Ms. Allegany County to a police officer and announce that they were under arrest for disorderly conduct. The trooper would escort the person back to the van.

Once the arrests began, Spike told Hope to begin moving people from nearby roadblocks to East Hill. As people began arriving, Spike set up another barricade fifty yards up the road.

At the original roadblock, the police filled the first van with seven demonstrators and were shutting the door to take them back to Belmont, when a man in his fifties wearing bib overalls stepped forward. "Don't take them. Arrest me instead," said Chuck Barnes, a landowner who had signed a waiver with the siting commission to let them on his property. Now, watching police arrest the demonstrators, he became troubled.

Four state troopers gently led him to the side of the road. "We can't arrest you, you've done nothing wrong."

Barnes thought to himself, "I'm an old Marine. I just don't like seeing somebody do something I ought to do." He stepped in front of the van, his arms outstretched.

"What is your name?"

"I'm Chuck Barnes, a landowner here." Extremely disturbed, he continued, "I had a brother die for this country and I have to stomach this? I've raised six kids on this farm. Do you think I'm just going to let them take it away from me? Not lying down I'm not."

"Mister Barnes," one of the state troopers gently said, "please get into the van. You're under arrest for disorderly conduct."

As the van pulled away, Chuck addressed the seven other arrestees. "I don't know any one of you. I've never seen you before. Thank you." He thought to himself, "When people you don't even know are willing to be arrested protecting your property, you have to do something."

Spike was standing beside the road, watching everything unfold exactly as he and everyone else had hoped. The police were arresting people slowly, one by one. "I've never seen half the people at this roadblock," Spike thought to himself, amazed that people actually showed up and did what they said they'd do.

A state trooper came toward him and he wondered if they were going to arrest him, even though he wasn't blocking the road. "Would you please move the truck that's behind us?" the trooper asked.

"There's no truck back there."

"Yes there is, and the driver won't move it."

Spike walked from the roadblock back toward the police cars and saw "Big Mura," a truck owned by a local farming family, pulled across the road. Danny Mura sat locked in the truck with the windows closed, defying requests by police to move his truck. Spike knew that part of the agreement with the state police was that ACNAG would not keep anyone from leaving the scene. So far the troopers had fulfilled their part of the bargain. They were wearing dress uniforms rather than combat gear and were arresting people in an orderly fashion.

"Danny," Spike yelled through the closed window, "move the truck."

As Danny pulled the truck out of the road, Spike walked back to the barricade. He noted that no further arrests were taking place.[5] The sheriff and Lieutenant McCole were talking, surrounded by media. "We don't seem to be getting anywhere," Scholes was saying, more for the benefit of the press than for communicating with the lieutenant. "There are twice as many protesters here as there were when we began the arrests. I see another blockade forming behind this one, and there are people pouring in to stand in front of it. I suggest that we regroup, consider what happened today, and have the siting commission decide how they want to proceed."

---

5. ACNAG leaders who later evaluated videotapes believed the police may have retreated in part at this point, because they were very disturbed at being blocked off. The ACNAG leaders' assessment that the police would never allow themselves to be blockaded would cause them to seriously miscalculate what the police would eventually do to get the siting commission on the land. See chapter 9.

"I agree that we can't get the technical team on the site today. We don't have the equipment to move the vehicles, even if we could arrest everyone."

"As soon as we arrest one person, more take their places," the sheriff concluded.

JANUARY 17, 1990

Mary Gardner, publicity director of CCAC, had been manager of the CCAC office in Belmont since September, the only person drawing a minimal salary to fight the nuclear dump. She had been at West Almond during the confrontation at Caneadea. The next day in the office, she saw Jim Lucey, who spent much of his time there when he wasn't lobbying in Albany. "I know," she said, "that Dave Seeger told us that officers of CCAC can't get arrested. I understand that, but I don't want to hide behind my position and not take a stand!"

"I know how you feel. Keeping them off the site is important right now, but there also has to be legislative action if we're going to win this thing. Seeger's right about keeping CCAC and ACNAG separate. Why not do support stuff when they come to West Almond?"

"I did that yesterday. I was stationed on the radio at the encampment, but I really want to stand in the road. What would you think if I decided to get arrested?"

"I'd respect you for sure," Jim gently responded. "Everyone's got to decide what's right for themselves. Civil disobedience is a matter of personal conscience."

"I know I'd have to resign as office manager."

"It'd be hard to replace you. You've done a great job. A lot of guys are willing to be arrested. Not many could do all the things you do around here."

"Okay, I'll think about it."

"Mary, I can't tell you what to do. You've got to do what feels right."

"Thanks, Jim." She immediately felt freed by his words. "If I decide to get arrested, just tell people that I'd already resigned."

"Good luck."

JANUARY 18, 1990 —

Two days after the state troopers arrested eight people at Caneadea, the technical team went to West Almond. Although the state police could transport people to the county jail in Belmont more quickly, since the site was only fifteen minutes away, it was also much easier to defend. Only three main roads allowed access, and farmers here were, for the most part, more militant than those at Caneadea and helped block the roads with huge tractors and other farm equipment. The Vigil had also focused attention on West Almond. Since the West Almond site was only fifteen miles away from Alfred University, activist students also took more of an interest in defending it.

On the day of the confrontation there were at least a hundred people at each of the three roadblocks; another two hundred people, gathered in the middle of the site, could easily be shifted to take the place of those who were arrested. Approximately half of the people were wearing orange armbands.

The police, however, were clearly committed to making massive arrests at West Almond. While no more than fourteen troopers had been seen at Caneadea, there were at least four dozen here. Several vans, serving as paddy wagons, lined the road.

The confrontation duplicated the ritual at Caneadea. Twice the sheriff and the technical team, followed by Tailgater's red truck and several press vehicles, arrived at a barricade and were turned back. At the third roadblock, the nearest one to the county jail, Bruce Goodale demanded that the police get the technical team on the site. State troopers moved deliberately and arrests occurred more quickly than at Caneadea. Both the protesters and Bruce Goodale, the head of the technical team, seemed more combative. Someone confronted him, "There's moral authority here. The moral authority of the people will eventually be heard."

He responded, "You have to understand we obey the law, and the law says New York State must have a nuclear waste facility. We're carrying out the law."

Gary Lloyd, who heard the exchange, asked him, "Do you feel what you are doing is ethical and responsible?"

"It's ethical and responsible to take care of nuclear waste. All of us depend on New York State's nuclear energy and New York must have a place to safely dispose the waste."

While people wearing orange armbands continued to be arrested in an orderly fashion, protesters surrounded Goodale, arguing that the waste was "low level" in name only and that a recent scientific study proved that low doses of radiation were three times more dangerous than previous governmental studies had shown. As the discussion became more heated, the sheriff intervened and told Goodale that he should get back into his car.

Mary Gardner was standing at the barricade, wearing an orange armband. She saw a man and woman in the center of the front line. They were Episcopal priests, stationed at the parish she had been attending in Wellsville. So far, neither had taken a role in the fight. She heard Lieutenant McCole

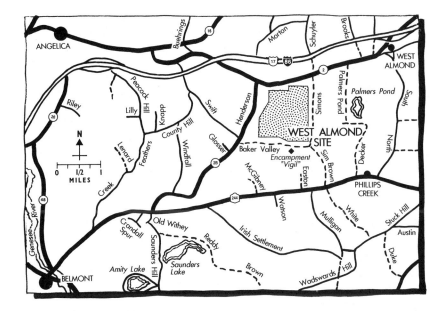

West Almond Site Map design by Craig Prophet

say, "Father Allegany County, you are under arrest, charged with disorderly conduct."

She heard a reporter talking to an architect who was standing in the road wearing a yellow armband. "I've decided not to be arrested today, but I'm proud to be supporting those who are." Just as he finished the sentence, Lieutenant McCole approached the man. "Will you move and let the technical team go on the site?" Mary laughed when the architect answered, "No, I'm sorry, but I can't do that." As he was being led away, she saw him tear off his yellow armband and drop it on the ground.

Mary felt extremely proud and knew she had made the right decision. The police were arresting people quite rapidly. She looked behind her, but could see only a wall of people, all wearing orange armbands, even though police had already arrested more than two dozen protesters. She was surprised when the man who was linking arms with her went limp after he was arrested. The troopers, prepared for every contingency, immediately brought out a stretcher and carried him back to the van.

Finally it was her turn. Lieutenant McCole introduced her to a female trooper. As she walked back to the police van, she saw Jim Lucey, smiling at her on the sidelines. "Great, sister!" he shouted. Other supporters clapped and she smiled at them, bent her elbow, and tentatively saluted them with a raised fist. When they got to the paddy wagon, she was told to place her hands on the van, while the trooper frisked her. She felt like a real criminal when the officer found a shotgun shell in the pocket of her dad's old hunting jacket. The trooper gently pushed her into the van.

At two o'clock Gary Lloyd moved fifty more people to the barricade. He saw Major Kelley come up to Lieutenant McCole and heard him ask how much longer it would take to get the technical team onto the site. Gary walked up to them and said, "Quite a few people are getting out of work at four o'clock. So there'll be a lot more people coming in; I just want you to know that."

"What do you mean?" Major Kelley asked.

"It's on the radio and I've talked to a lot of people who told me that they'd come over as soon as they got off work. I wanted you to know, because they might clog the road behind you. We'll have to make routes for them to enter."

Kelley and McCole walked back toward their cars, while another person was arrested. For the first time, a police helicopter buzzed the site, apparently trying to estimate the number of people. Abruptly at 2:30 the arrests stopped. Thirty-one people had been arrested. When questioned by reporters, McCole told them that they gave up, because they "weren't accomplishing anything."

The sheriff told reporters, "As one person was arrested there was another, maybe two" replacements, "and up the road, there were many more" who were ready to take their places.

Angelo Orazio, the chairman of the siting commission, communicated his frustration to the press a couple of days after the technical team returned to Albany. He blamed the "leaders" of the protest movement for disseminating false information. "We have to remember that in spite of our efforts many people" in Allegany County "are suffering from information starvation." He claimed that people were at the sites out of curiosity and were not involved in civil disobedience. Leaders of this "fringe group" are treating people in Allegany county "the way you feed mushrooms; you keep them in the dark and feed them manure."

No comment more exasperated, angered, and paradoxically elated people in the county than Orazio's comparing them to "mushrooms." An editorial in the *Olean Times Herald* argued that the people blocking the sites were not the "fringe," but the "heart" of the county: "Let us look at those arrested this past week: two ministers, both in their fifties; an architect; several farmers; several

business people; teachers; a town clerk; a high school student; college professors; stay-at-home mothers."[6]

Orazio's mushroom comment re-opened Gary Ostrower's public feud
with the siting commission, which had intensified a month earlier. In December the siting commission had unsuccessfully tried to establish a dialogue
with a few selected scientists and engineers at Alfred University. Ostrower
asked through the press why they wanted to initiate a dialogue with some new
scientists when they had never responded to the report that his committee of
scientists had given them nine months earlier: "Days turn into weeks, weeks
turn into months, and still Allegany County receives no straight answers. We
gave you our report last March; since then, we've gotten little from your
commission other than public relations and court orders. You owe us an apol-
ogy, not cheap criticism."

Ostrower found it ironic that the commission accused the protesters of
treating people like mushrooms. Not only had they failed to respond to the
scientific report, but they had refused to supply the software to read the
"Geographic Information Survey" data file that a court had ordered them to
give to the county. Ostrower now wrote, "Showing contempt for the scientific method, the commission has refused to make the data available for
scrutiny by independent scientists." He concluded his open letter: "So be it.
Mr. Orazio should remember . . . that mushrooms are hardy. They grow
even in fields like those in Allen and Caneadea and West Almond, and they
keep coming back."

Dozens of letters from all sorts of people flooded the newspapers in response to Orazio's remarks. Farmers and professors, homemakers and merchants wrote letters. They told Orazio that they had spent considerable time
reading about the storage of nuclear waste and they resented his arrogance in
suggesting that they were mindlessly following leaders who were keeping
them in the dark. Peg Jefferds, a homemaker, wrote, "I am a member of
ACNAG. I do my own research. With access to the pros and cons of nuclear
waste, I feel capable of making my own decisions."

---

6. Of the thirty-nine people arrested at both Caneadea and West Almond, there was
also a diversity of age and geographical area. Three people were in their sixties, four in
their fifties, eleven in their forties, sixteen in their thirties, four in their twenties, one in
late teens. Six people were arrested from Belfast; five each from Belmont, Alfred, and
Wellsville; four from Almond; three from Andover; two from Cuba; and one each from
Arkport, Angelica, Dalton, Fillmore, Little, Genessee, Middlesex, Caneadea, Allentown,
and Naples. Geographically these areas represented all parts of the county and a few
places in nearby counties.

Bernard Piersma, a professor of physical chemistry at Houghton College, had not been a member of Ostrower's committee and was no leader in the movement. He presented his credentials in an open letter to Orazio. He had received his Ph.D. from an Ivy League college, had studied on postdoctoral fellowships from the Air Force Office of Scientific Research, and had taught college-level courses in nuclear chemistry and on the biological effects of radionuclides. After discussing several scientific issues, Piersma wrote, "I certainly agree with Mr. Orazio that the people of Allegany County are being treated like mushrooms. Where we do not agree is who is treating us that way. Even the name 'low-level radioactive waste' is a calculated deception since the materials to be disposed of are neither low in quantity nor low in radioactivity."

Kay Eicher, a high school teacher, whose house had served as communications center for the action in Caneadea, took a more whimsical approach. She thanked the siting commission for helping ACNAG recruit members. When Bruce Goodale "called us decent but ignorant," implying that "we were mindless duds being manipulated by a small group of 'protest leaders,'" recruitment went up." She also thanked Orazio for his mushroom comment. "'Keep up the good work'"; we wouldn't have been able to afford such "a fantastic public relations campaign!"You may be able to "turn the whole county (perhaps the whole southern tier) into one large mushroom patch!"

Most people, however, deeply resented Orazio's comments. A professor of German at Alfred University wrote, "Mr. Orazio, I want to assure you that I am no 'leader' in this protest movement," but I "find your comments insulting and arrogant" because they "imply that I am easily misled, gullible and foolish." They also "imply that I haven't . . . read the many reports . . . on radioactive waste, . . . that my thoughts and actions" are "controlled somehow by rhetoric of 'leaders,'" and "that you hold the 'facts' about radioactive waste disposal and that any conclusions other than your own are invalid."

When representatives from various constituencies met in John Hasper's Belfast office to discuss coordinating the county's anti-dump activities, the state assemblyman reported that he thought it might be possible to get Orazio replaced as head of the siting commission.

"Far out! That would be a victory," exclaimed Jim Lucey.

Nearly everyone nodded their heads and grinned. Scott Weaver, a geologist at Alfred University who had co-authored a devastating part of the county's scientific report on the commission's use of statistical data, however, wondered aloud, "Are we so sure that we want to have Orazio replaced? He seems to be doing a great job for us." Everyone laughed, and Jim immediately revised his earlier inclination to go after Orazio. Stuart Campbell thought of Clemenceau's observation that the only thing worse than a bad priest is a good priest, but remained silent. Ted Taylor concluded the discussion, "We certainly shouldn't expend any effort in getting him replaced."

In order to nourish Orazio's "informationally starved" people of Allegany County, the siting commission intensified its efforts to find an office in the county. Property owners, many of whom badly needed the money, routinely refused to rent their storefronts. Most agreed with the strategy of noncooperation. A few worried that their neighbors would see them as collaborators.

In early February, for example, two Wellsville businessmen, John Barnett and Gary Fuller, who were financially strapped for funds entered into serious negotiations with the siting commission when they were offered one thousand dollars a month plus insurance costs and a six-month guaranteed lease for a space that normally commanded between three hundred and five hundred dollars per month according to the town assessor. This amount of money would allow the two men to get seed money to fulfill a dream of starting a new business venture.

Word somehow leaked out before the contract was signed and five or six anti-dump activists visited the men and argued that renting their office would give the siting commission its first foothold in the county and could damage the movement. They told the two men about a meeting, chaired by Assemblyman John Hasper, with representatives from many county groups—CCAC, ACNAG, the county legislature, the Soil and Water Conservation District, and selected scientists—where everyone agreed that any cooperation with the siting commission would only help them site the dump.

The businessmen weren't entirely convinced and argued that an office might actually benefit the movement, becoming a fixed site for picketing and protest. Unsettled by the arguments of the activists, however, the two men sought advice from other business people in Wellsville. All counseled caution, telling them that in the long run public animosity would outlast short-term profits.

Two days later, with the press hovering around, the two men announced that they would not rent their office to the siting commission. They told the press, "We want to make it clear that we're anti-nuclear. As business people we have a lot of friends and business associates and we don't want to hurt anyone's cause." One of the men explained, "At this point, they [the citizens] are winning, 100 percent. To allow the siting commission to come in would be a blow. There's no neutral ground" in Allegany County. They insisted that they were not intimidated by anyone. One of the men asserted, "I'm the type of person who, if someone says to me, 'You can't' or 'You won't,'" I say, 'I can' and 'I will.'"

One of protesters who had been involved in discussions with the two men told reporters that "their decision is for the good of the entire county. They've made a considerable sacrifice and it shouldn't be forgotten. They've shown themselves to be exemplary friends of this community and its children."

By the end of February the siting commission admitted that they had been unable to rent space anywhere in Allegany County and announced that they would bring in their own mobile office and spend two weeks every month touring around the county. They would begin by spending one day at five different locations between Tuesday, March 6, and Saturday, March 10.

TUESDAY, FEBRUARY 27, 1990—

A small group of ACNAG leaders met in Gary Lloyd's basement to decide whether they should call a steering committee meeting to plan a civil disobedience action to stop the siting commission's van—a recreational vehicle.

"The two landlords in Wellsville, a woman who owned a cafe in Belfast, and many others sacrificed a lot of money to keep the siting commission out of the county," said Stuart Campbell. "We shouldn't let them just waltz in here without a fight. I don't have any great ideas, but we should call a steering committee meeting to discuss it." He paused. "You told me yesterday that you don't think we should get involved, Tom. Why don't you present your argument."

I seemed destined to cast myself into the role of devil's advocate at these meetings! "Everyone in the room agrees with the strategy of noncooperation," I began slowly. "It's not only stupid to play the siting commission's game, but . . ."

"It's like a sheep going up to a wolf and trying to explain why he should become a vegetarian," Spike interrupted. "The siting commission can just mark their check list with an X next to 'propaganda.'"

"Actually, I think they call it 'public outreach,'" I said. "No need to convince anyone here, Spike, but the question is whether ACNAG should mobilize everyone to stop a van that nobody's going to pay any attention to anyway."

"You're wrong about that," said Gary Lloyd. "Sending the van in here will anger lots of people."

"That's not what I meant. Sure, people are going to get riled up, I just don't think that anyone will seek information from them. I don't think that they'll accomplish their purpose."

"But they'll still be able to check it off their list," said Spike. No one said anything for a minute. Spike's argument had almost convinced me that ACNAG would have to take a stand against the van.

"Of course they will," I tentatively responded. "But they checked the precharacterization of Allen off their list after being on Giovanniello's land for only twenty minutes. That just made them look ridiculous. We're keeping them off the sites. That's critical. This isn't worth the effort."

"This time I agree with Tom," said Sally Campbell. "This is small potatoes. Everyone's going to resist anyway. You can't send the siting commission into a nest of hornets without getting stung."

"Maybe that's why we should get involved—to keep people from getting hurt and make sure that everything goes smoothly," said Gary.

"It's not our job to be cops," Spike said. "ACNAG wasn't formed to protect the siting commission. If they want to be fools, that's their business."

"I don't know if everyone really knows how much effort it takes to mobilize the phone tree before each action," said Hope Zaccagni. "Not only do I have to check to make sure that all the links are in place, but we have to make the message very simple or it'll get garbled. The van is going to be at five different locations in five days and it would be difficult to have a simple message."

"One thing we don't want to do," said Gary, "is plan some half-assed action. We've gained people's respect because everything we do is well planned."

"Is it reasonable to organize five different actions in the next week?" I asked.

"You've all convinced me," Stuart said. "I know how much work it was to plan the defense of the sites. I also remember organizing the Belmont action. . . ."

"I think that's exactly what I'm feeling," I interrupted. "If we block a van that's coming into the county, it will be a step backward. We've got to keep focused on protecting the sites and not get sidetracked. This is a no-win situation. If our plan didn't work, we'd look stupid; if it did, the siting commission would say that protest leaders are keeping information from the people."

No one spoke for a minute. It was clear that there was a consensus. Sally finally broke the silence. "I'm tired of dealing with the media. Since you, Tom, feel so strongly about this, why don't you write the press release and call the newspapers this time."

As always, the price for winning an argument in ACNAG meetings meant more work. On March 1, the *Wellsville Daily Reporter* highlighted ACNAG's position in an article reporting the schedule for the siting commission's mobile office: "Tom Peterson of Alfred, an ACNAG spokesperson, issued the following statement Wednesday: 'At this time ACNAG has no plans for any organized action to deal with the siting commission's mobile propaganda effort. As always, individuals are free to follow their own consciences. . . . ACNAG feels its primary responsibility is to keep the siting commission off the sites.'"

CCAC issued a press release a day later. They would bring their own mobile "Mushroom Information Center" to places where the van would be. They also discouraged people from entering into a dialogue with the siting

The Siting Commission's Van: Skunked © Steve Myers 1990; all rights reserved.

commission: "People should be aware that any conversation can and will be used against us by the commission to bolster the impression that it is obtaining the required public participation." They also cautioned protesters to remain nonviolent: "Anyone acting in a violent manner is no part of CCAC."

❖

TUESDAY, MARCH 6, THROUGH SATURDAY, MARCH 10, 1990—

ACNAG's decision left a vacuum that others eagerly filled. In each of the scheduled communities, protesters organized specific activities to harass the siting commission. On the first day in Wellsville protesters occupied every parking place in the lot where the commission had a permit. Wellsville's mayor explained to the press that the van was not allowed to park on Main Street, because he and the fire chief did not think it would be safe. "Obviously," he said, "people from the Wellsville area and the county didn't want them here. . . . It is a public parking lot and I guess the public got there before the commission did."

The second day in Belmont a tractor pulling a manure spreader blocked the van in its parking place, while B.A.N.D.I.T.S. music blared into the vehicle. Protesters held up signs asking people to honk their horns if they wanted the siting commission to leave the county. Almost everyone honked, including several tractor-trailers. One of the men in the van later told reporters that they had seen three people that day, only one of whom took any information. One person went into the van with his feet covered with pig manure. "Soap and water can clean up the mess I'm making here," he told them, "but cleaning up the shit you want to bring into the county won't be so easy."

On the third day the mobile office paid an unexpected visit to Canaseraga, a tiny village on the northeast edge of Allegany County. Although no one had organized a protest there, an angry crowd of residents yelled and shouted for the van to get out of town. One man repeatedly slammed the van's door and angrily shouted obscenities at the men inside until other protesters calmed him down. In Fillmore on the fourth day, so many protesters' vehicles lined Main Street that the van had to park on a side street outside the village limits.

The van's final day at Almond was clearly the wildest. Here, on a Saturday when people were off work, the siting commission arrived in the most hostile area of the county, where Steve and Betsy Myers had begun organizing more than a year earlier. Tensions had been building all week and this was the last chance to demonstrate against the dump, since the mobile office would be returning to Albany that afternoon.

The protest in Almond had a very different tone than the nonviolent resistance at the sites, where protesters had carefully choreographed their activity with military-like precision. Here the mood was more like Mardi Gras in New Orleans. Since there was no central planning, individuals created their own unique theater of protest.

Approximately four hundred people were out on the streets of Almond, and cars lined both sides of the mile-long county road that wended its way through the valley, defining the village's backbone. The siting commission's van drove up the road, seeking a parking place, turned around when it reached the edge of town, and started back down the street. Three dozen demonstrators stepped in front of it, stopping it dead in its tracks. More people arrived, some of them plastering the vehicle with antinuke posters and bumper stickers.

Stuart Campbell was observing the activity from the sidelines and found it strange that no cops were present, even though Almond was a regional substation of the state police. When people began to rock the van, he worried that the men inside would panic and start driving through the crowd, or that the people might even tip it over. Stepping into the crowd, he persuaded those in front to open up a space so the citizens could escort the vehicle out of town.

In the half-hour that it took the van to go back through the village, people performed their individual acts of street theater. Three people, dressed in white sanitation-worker uniforms, brought their cardboard geiger counters and checked for radiation. A man dressed as the skeleton of death jumped in front of the procession to lead what was becoming a festive parade. An effigy with the tags "siting commission" and "Mario Cuomo" was tied to the back bumper and dragged through the street. Someone else attached a box of manure with papier maché mushrooms "growing" out of it.

Sue Beckhorn had brought a dead skunk she'd found frozen in the ice near her home. She decided to put it in the box of manure. Walt Franklin was wearing a Seneca Indian corn mask as he walked beside the van, now "tarred" with antinuclear slogans and "feathered" with orange crepe paper. He would never wear such a mask lightly, but felt that the Seneca spirits would approve of its becoming part of a protest to keep nuclear trash out of the region. When he saw Sue put the skunk in the box of manure, he thought he knew a better place for it, picked it up, and stuffed it in the van's ventilator.

The vehicle finally made it to the expressway ramp. People cheered and hurled a few rotten eggs as it sped off toward Albany.

The siting commission apparently concluded that they would not be able to convince the people to voluntarily accept a nuclear dump. They never opened up an office, nor brought their van back into Allegany county. Since they couldn't persuade the people, they would now use the brute force of the state troopers and a decisive injunction of a New York state court.

8 Outside the Law

If the protesters continue to react as they have to date, it would be impossible for the State Police to protect the Siting Commission's representatives without becoming, in effect, an occupying army in Allegany County.

—Lieutenant Walter Delap, State Police,
in court deposition, February 5, 1997

FEBRUARY 20, 1990—

SPECTATORS IN A PACKED state Supreme Court room in Buffalo were anxiously listening to Jerry Fowler try to convince Judge Jerome Gorski not to issue an injunction restraining protesters from interfering with the siting commission's work. Specifically named in the injunction were Gary Lloyd, Stuart Campbell, Sally Campbell, Spike Jones, and the thirty-nine protesters who had been arrested in January at Caneadea and West Almond. Even more troublesome, the proposed injunction also named John Doe and Jane Roe—in effect, anyone hindering the siting commission's work.[1]

Peter Sullivan, assistant attorney general for New York, explained why he was once more seeking an injunction. Central to his case were affidavits of three state police officers. The troopers swore that they had made serious efforts to get the technical team onto the sites, but estimated that they would have needed one hundred fifty to two hundred troopers to have accomplished their mission.

---

1. The siting commission had tried to get an injunction from Judge Gorski earlier in December. The judge had then declined to issue an injunction, ruling that the state had not shown that the regular criminal procedure was ineffective. Jerry Fowler had then argued that the district attorney could charge the protesters with "Obstruction of Governmental Administration," which was a misdemeanor with a maximum punishment of six months in jail and/or a one thousand dollar fine.

At West Almond they would have had to arrest at least four hundred people—and that was a "best case scenario." If protesters decided not to cooperate with the police and had to be carried away, many more troopers would have been needed—indeed it would have taken "an occupying army." Sure, the police were "committed to enforcing the law, . . . but the court should realize the limitations." Fifty-seven troopers along with "several vehicles, a mobile command center, a helicopter, and three prisoner buses" had been involved at West Almond. The use of more troopers would seriously have strained their resources, they argued. Even more troubling, they said, such an operation would harm "our relationships with the citizens," thereby destroying "our ability to effectively provide law enforcement in the county for years to come."

Sullivan concluded by arguing that the criminal law was clearly not working. The police could not get the technical team on the site; the Allegany County district attorney only charged the protesters with violations, and local judges imposed meager fines that ranged from twenty-seven dollars, the minimum state-mandated "court tax" to thirty-two dollars. An injunction would make defiance of the law less successful by making it a civil matter.

But why? The judge, the lawyers, and the spectators in the courtroom all understood the implications since the head of the siting commission had explained them to the press a couple of weeks earlier. Those violating the injunction would not have the right to a jury trial. No longer would local judges impose the fines; the judge in Buffalo who issued the injunction would handle the matter. The judge, Orazio said, will be able to "levy fines substantially greater than those levied by a justice of the peace on a disorderly conduct charge." Large fines and possible jail terms will "give adequate incentive to protesters to stop interfering with our work."

Knowing it would be an uphill fight, Jerry Fowler had spent many hours preparing arguments. The protests in Allegany County were at a critical juncture. If he could not convince Judge Gorski to deny the state's request, protesters might become more reluctant to engage in civil disobedience. Not only would a one thousand dollar fine seem enormous to Allegany County people but the monetary threat would actually be much greater—the judge could impose "damages" for extra costs to the state. Fowler wondered how much it cost the state to pay sixty police officers overtime and bring in a helicopter, a mobile office, and buses. The siting commission would surely claim that every day's delay cost the state huge amounts of money. How would protesters react to the threat of losing their cars and homes? Would they continue to defy the state and act outside the law?

Fowler had asked ACNAG to send as many people as possible to this hearing, even though Buffalo is nearly one hundred miles away. He wanted to convince the judge that the citizens were vehemently opposed to the nuclear dump and that an injunction would create more tension without solving the state's predicament. Before a packed courtroom, he argued the case in his mild and measured manner that had been so successful in the rural courts of Allegany County. "An injunction won't intimidate the protesters. The siting commission doesn't understand what's happening in Allegany County, and until they do, issuing an injunction will just create one dangerous situation after another. It won't stop the protests."

Fowler then referred to a deposition in which Sheriff Scholes unambiguously stated that an injunction wouldn't work. In fact it could even make the situation more dangerous. Fowler suggested that the state had not pursued criminal remedies vigorously enough. None of the four people named in the December draft of the injunction had yet been charged with a misdemeanor. Nor had any agency of the state contacted the county district attorney to suggest that he become more aggressive.

Sullivan, whose combative style contrasted with Fowler's folksy one, jumped from his seat, sputtering, "That argument's a canard and an attempt to mislead the court. The state police have declared that the criminal laws won't work."

Fowler responded, "The siting commission and the police have only begun to explore the criminal law."

The judge then asked Fowler, "Does a five dollar fine have any chilling effect?"

"I don't think one thousand dollars fines would chill this. There are people quite willing to suffer jail sentences." Fowler paused and when no one spoke, he added wryly, "I've actually been berated by a few citizens because they haven't yet been incarcerated."

The spectators in the courtroom chuckled. Until this point the citizens of the rural county had been extremely quiet, perhaps a bit awed by the lavish mahogany courtroom, which marked the wealth, power, and pretensions of the Empire State at the turn of the century. Fowler continued, "The people in Allegany County are extremely upset. They regard the siting commission's proposal as an engineering experiment. They can't show us that a nuclear dump will not leak. Without an injunction, the commission will have time to do some tests, and make some prototypes that will make it easier to sell. The siting commission has failed to convince the people. Its public relations activities have been dismal."

"What can this court do about the commission's failure to communicate with the people?" the judge interjected.

"Don't issue the injunction," Fowler replied. "That would force the siting commission to do the necessary, preliminary work of testing various designs and communicating with the people."

"Testing is exactly what the siting commission is trying to do," Sullivan said sarcastically. "They are under a stringent timeframe that has been set by federal law. Mister Fowler is acting as if the commission created its own legal obligations. It doesn't want to do this; it has to do this. He's indirectly asking the court to enjoin the commission from going through the siting process. It's ludicrous."

Visibly agitated for the first time, Fowler repeated, "An injunction will not solve any of the siting commission's problems. The technical team will then confront the same—even larger—blockades. They'll have one dangerous situation after another." Then, looking directly at Sullivan, Fowler injected his own sarcastic comment: "Of course, they can build the dump in Allegany County, but they'll need a security force with over five hundred officers, its own jail to hold at least two hundred people, its own prosecutor, and its own wrecking equipment to tow vehicles away. If they're willing to do that they can put it in Allegany County. That's the price they'd have to pay."

Judge Gorski was not going to allow the two attorneys to continue their wrangling, and forcefully interjected, "If there are no further arguments, I will reserve judgment. You may expect my decision in a few days." No one spoke. "Court is adjourned."

A few days later the judge issued a blanket injunction, not only against the forty-three named protesters but against anyone interfering with the activities of the siting commission. Reporters asked ACNAG leaders whether the civil disobedience that had kept the technical team off the sites could be sustained. The leaders remained publicly confident, but privately they worried.

MARCH 27, 1990—

Sally Campbell, Gary Lloyd, and I joined Larry Scholes and Bill Timberlake in a conference room in the county courthouse. As soon as the siting commission announced that it would come into the county to walk over the Caneadea site on April 5, the sheriff asked to meet with a few leaders in ACNAG.

"I don't have a lot of information about what'll happen next Thursday, but I thought we ought to touch base," the sheriff began. "The injunction has changed things. We won't have to arrest people to enforce the injunction. The attorney general's office will have photographers at the roadblocks and will take pictures of people blocking the road. I won't have a choice. I'll have to identify anyone I recognize."

"Does this mean that the troopers aren't going to arrest anyone this time?" asked Sally Campbell.

"I didn't say that. I think that we'll pretty much do what we did before. If I find the road blocked, I'll have to call in the state police. The state police will probably arrest people who're blocking the road, and we'll charge them with disorderly conduct. I assume that the attorney general's office will later charge these people with disobeying the injunction. The police will have videotapes of other people who're there. If identified, they could just as easily be charged. I assume they'd love to get some of the people specifically named in the injunction. I know a lot of you, and if I can recognize anybody, I'll have to do that."

"Of course, it's still pretty cold, especially for March," interjected Timberlake. "Lots of people wear ski masks around here in winter. If they're wearing them, we won't be able to tell who they are."

"A few people in ACNAG have wanted to dress up like beavers, foxes, and other animals in the county to make the point that the dump also threatens the wildlife," I said. "Most of the group thought it wouldn't be dignified, but maybe we'll have to reconsider wearing animal masks."

"Jeez! Whatever you guys do, don't dress up like Mickey Mouse!" Scholes joked. "I'd have a hard time keeping a straight face."

Everyone laughed and the tension eased a bit. Sally thought to herself that the two law enforcement officers seemed uncharacteristically grim. None of the other meetings with the two men were lighthearted, but there hadn't been the tension that she felt in this meeting. The sheriff now seemed to be making a conscious effort to relax everything a bit. She remembered him joking about being Andy of Mayberry on a bad day when he had first encountered ACNAG protesters surrounding the siting commission's car in Belmont. Now Larry seemed loosely tethered, Bill tightly wound. They're worried, she thought. They know something that they're not saying.

"Are we positive that the siting commission will be coming to Caneadea?" she asked out loud.

"Yes," the sheriff answered. "The technical team won't be playing games going from one blockade to another. And they're going to take the most direct route—across the bridge at Caneadea and then straight up East Hill Road to the site."

Gary Lloyd didn't trust any information that came from the siting commission." What makes you so sure that the siting commission isn't lying?" he asked.

"The information's from the state police and they've never lied to me before. I made it pretty clear after Allen that I'm not going to have people chasing the siting commission all over the county. That's just too dangerous. We're lucky there hasn't been a serious accident on these icy roads. If they want my cooperation, they won't lie to me. I've told them that they'd better inform me in advance of any change in plans."

"We've been told that there'll be even more state police this time," interjected Timberlake. "Now that they've got their injunction, they're really going to try to get onto the site, or at least arrest lots of people."

Sally said, "We've been told by someone close to the attorney general's office that Gorski wants the police to arrest at least fifty people. He'll give them the maximum fine. He thinks that'll break the movement."

"I know nothing about that," said the sheriff, "but it seems reasonable."

"We had lots of farm equipment blocking the road behind the protesters at West Almond," I said. "We wouldn't necessarily need so many people blocking the road. They'd have a hard time removing all that equipment. I'm not sure that a tow truck could easily move those huge tractors." I smiled. "Most tow trucks in the county were broken down during the last actions."

"If you didn't have any people blocking the road, then what would prevent the police and siting commission from going around the barricades and marching onto the site?" Timberlake asked.

Stunned silence! No one could answer the question. The three leaders of ACNAG retreated into their own thoughts. This was a "Catch-22." If the protesters used their bodies to blockade the site, they'd be arrested and Gorski could fine them one thousand dollars each—even more for damages. If they tried to use equipment, then the technical team would just walk past them to the site.

No one spoke for at least a minute as the tension mounted. Finally I said the first thing that came to my mind. "You know that we're determined to keep the siting commission off the land at all costs. I guess we'd have to go back to the tactic we used at Allen. We'd have to surround the team when they got close to the site."

"You understand what that'd mean?" Timberlake asked rhetorically. "I've been to lots of demonstrations when I was a cop in New York." He stood up and touched Gary and me on the shoulder. "Stand up and link arms." When we complied, the undersheriff raised his arm and slowly brought it down between our linked arms. The pressure made it difficult to stay linked. "Now imagine that I'm swinging a billy club at full force. The state police won't let you keep the technical team prisoners. They'll play rough if necessary. You'd better think through all this!"

The undersheriff was just now clarifying for himself and for everyone else the risky scenarios that the injunction had unleashed. The sheriff had already lost several nights' sleep thinking about the next action. When he couldn't sleep, he'd get up and stare out his window into the darkness, considering all the possibilities.

No one spoke. Sally thought to herself, "They're going to march onto the site. This is what the state police have told the sheriff. This is what Timberlake is trying to tell us."

The sheriff realized that ACNAG hadn't considered the alternatives. He'd hoped to learn more about their plans, but realized that they hadn't formed any. And violence, he knew, was more likely when people are taken by surprise.

The sheriff now tried to reassure the protesters as best he could. "Neither Bill nor I know exactly what's going to happen. I've been told to have a deputy ready to read the injunction to any people blocking the road. The attorney general's office wants to be sure that people who're arrested know there's an injunction against interfering with the technical team. Since an injunction is a civil matter, I'll be in charge of whatever happens, at least until I'm forced to call in the state police. I'm pretty sure that McCole will be in charge, so everything should remain peaceful as long as everyone keeps their cool and no one panics. I assume you'll have spokespeople at the roadblocks as you've done in the past. I'll be as clear as I can about what's happening. Anything you can tell me ahead of time might lessen the chance of violence."

"We're having a steering committee meeting tomorrow night," said Gary. "We'll be exploring our options then. We've got a lot of things to think about."

MARCH 28, 1990—

I watched people file into the classroom on the second floor of Kanakadea Hall on the Alfred University campus. The badly worn, moveable desk-chairs, their tops carved with the initials of generations of students, contrasted with the room's elegant architecture of a bygone era. Six large, slender windows with decorative casings and a pressed-tin ceiling with a geometric design attested to the importance that people of the previous century gave to education. The smell of the freshly oiled hardwood floor triggered memories of my own elementary school.

Waiting for everyone to arrive, I thought about earlier strategy sessions. This group had come up with creative responses to the ever-increasing pressure from the state. Most people left their egos at the door as they focused on defeating the state's plans. Individuals felt free to speak their minds, but they also listened carefully to others. Until now we had achieved consensus on all decisions—a remarkable feat for a steering committee of more than forty people. The injunction changed the situation considerably. Could a group that was becoming larger at every meeting still make decisions by consensus, especially in a time of crisis? Fifty people had filled the room by 7 P.M. and more people crowded into the space, perching on windowsills and leaning against the walls.

I was as startled as everyone else when four people filed into the room wearing masks. Sally Campbell had transformed herself into a cat. Spike, wearing a cowboy hat and bandanna across his mouth, looked like a robber in a B-grade western. Stuart Campbell wore some cheap Halloween mask, bought from the dollar store in nearby Hornell. True to form, Gary Lloyd had carefully prepared for his part, wearing the costume of a mountain man with feathers in his hat, a blackened face, and buckskin pants.

Rather than the expected laughter, ACNAG's leaders faced the palpable silence of disapproval. They seemed to be advocating that the next action would take on the appearance of Mardi Gras. Feeling the people's resistance, Spike and Stuart quickly removed their masks. Using such words as "essential," "imperative," "mandatory," and "critical," Sally assertively explained why everyone would have to wear masks at the next demonstration. Gary argued that everyone could choose masks that truly represented the animals and people of Allegany County.

Immediately comments began flying around the room. "So far, our actions have been dignified. I just don't like the thought of turning ourselves into clowns."

"The press and people in the county have supported us because we've been serious and acted with class. What kind of impression will we make wearing Halloween masks?"

"I loved tarring and feathering the siting commission's van in Almond. The theatre was great. But even that activity verged on violence, because everyone was doing their own thing. Creating this kind of atmosphere at the sites would be very dangerous."

Sally again became quite impassioned. "We really don't have a choice. The injunction has closed off a lot of our options. We can't just roll over and fall into Judge Gorski's trap. We have to focus on keeping the technical team off the sites and do what's necessary."

I had moderated an ACNAG steering committee meeting a few months earlier about civil disobedience and tactics. After the meeting people thanked me for running it, and several suggested that I moderate future ones. They claimed that I allowed everyone to speak but also sensed when there was general agreement on a point and was able to move the meeting along. Tonight I saw everything becoming chaotic.

I stepped forward and moved to the center of the room. "The issue of whether or not to wear masks is perhaps the most important thing we need to discuss," I began, claiming authority to moderate the meeting.

"We really don't have a choice," Sally interrupted, "unless we want to give up the fight. How many people in Allegany County will do civil disobedience after they're fined one thousand dollars. Even worse, the judge could assess damages on top of that."

"Sally, I know you feel very strongly that this is the only solution. You may be right, but you're going to have to back off and let everyone discuss the pros and cons." She looked at me and shrugged, realizing that she'd have to let people reach their own conclusions.

"First, let me be sure everyone understands what the injunction means," I began. "In the last actions the police arrested us for blocking the road. We were charged with disorderly conduct and taken before Allegany County judges who gave us minimal fines. The D.A. could've charged us with a misdemeanor, but then we'd have had the right to a jury trial and he knew he'd have a difficult time getting a conviction. So he agreed to charge us with a violation and to recommend minimal fines if we agreed to plead guilty to the lesser charge and not tie up the courts.

"If we're charged with violating the injunction, we'd have no right to a jury trial and we'd be brought before the same Buffalo judge who issued the injunction. He'll pass judgment on our guilt and there's every reason to believe that he'll fine us one thousand dollars and give us six months in jail. Someone in the attorney general's office told us on the q.t. that the judge wanted the police to arrest fifty people, so he could break the movement."

"I don't see what any of this has to do with wearing masks," someone said. "Whether we're wearing masks or not, we'd still be arrested and could be charged with violating the injunction."

I felt that I'd already talked too much. One thing I knew about running meetings is that the moderator should only be minimally involved in the discussion to preserve the sense of neutrality. "Who wants to explain why wearing masks would make a difference?"

"Sally, Tom, and I met with the sheriff yesterday," Gary began. "He told us that there'd be police photographers at the roadblocks. Even if people aren't arrested but are near the action, they could then later be identified and charged with violating the injunction. The sheriff said he'd have to cooperate and identify people if he could."

"Let me back up a little," Sally interjected. She was now calm. "Remember how we had farm equipment and cars blocking the roads behind us at West Almond? We knew it'd take the police a long time to get rid of the mess and we'd have time to regroup and bring in more people. This time, because of the injunction, we might have to rely more on blocking the road with cars and tractors, because we don't want a lot of people arrested. But that's not going to do us any good if the police are taking our pictures."

"I don't think anyone has really addressed my objection to wearing masks," Diana Zweygardt interjected. A weaver and fabric artist in her mid-forties, she had been active in opposing the dump from the beginning of the struggle. She and her husband Glenn had been founders of the Alfred chapter of CCAC and both had participated in organizing the opposition to the dump in Betsy and Steve Myers' kitchen. Diana had spent many hours contacting groups that were fighting nuclear dumps in other states and had compiled newspaper clippings of their activities. Her efforts helped link CCAC and ACNAG to antinuclear protesters in other states. She had joined ACNAG because she supported civil disobedience as both a moral and effective means of protest.

"We've been strong so far," she said, "because we've been willing to commit ourselves to fighting against injustice. We had moral authority, because we were willing to accept the consequences of our actions and pay the price. Nonviolent resistance depends on being open. If we now decide to hide behind masks, we'll have lost all our moral authority."

Most of the people in ACNAG were pragmatists. They supported civil disobedience because it exposed the siting commission's arbitrary use of power. Even better, it had been effective in keeping the technical team off the sites without violence. Almost everyone in the room was willing to risk jail and huge fines if that were the only recourse, but they wanted to hear more about the alternatives. I could see the gulf between the pragmatists and c.d. purists widening as positions began to grow more rigid.

"Let's examine each of the arguments against wearing masks," I said, noticing that people had begun fidgeting. "Here's what I'm hearing. Quite a few people don't like the idea that we're not facing arrest openly like we did in the past. As many of you know, we've got a sizable arrest fund. An anonymous donor started it with five thousand dollars and it's grown to nearly ten thousand dollars in only a week's time, but that clearly won't go very far if forty or fifty people are arrested. Diana is suggesting that we should just thumb our noses at the judge and siting commission and take the consequences, because that's the only way we can maintain our integrity. How do the rest of you feel?"

I called on Mary Gardner, who, since her arrest in West Almond, had committed herself completely to ACNAG's nonviolent resistance. "I don't like the idea of hiding behind masks. I've always liked the fact that we've been open and willing to be arrested. But this injunction is an abuse of power. The state isn't willing to try us in an ordinary court of law. Isn't that what the American system of justice is about? A jury trial by one's peers? In a sense the judge has put us outside the law—at least outside the system of American justice."

"Is there any way we could limit the number of people arrested?" someone wondered. "We could have a few people blocking the road without masks. We could choose people who're really respectable—people the police would rather not arrest."

"You mean, not riff-raff like Spike and Stuart," someone else joked. The original antagonism against leaders imposing their will on everyone dissipated in the good-natured humor.

"That's a great idea! What if most of us wear masks, but we have a few prominent people standing openly in the road—a doctor, a farmer, a businessman, a teacher."

"Yeah, let's not let them decide who they're going to arrest. Let's us decide."

Roland Warren, a tall, quiet-spoken seventy-four-year old, raised his hand. I guessed that he might side with Diana Zweygardt and argue for principle over pragmatics. A sociologist who had taught more than twenty years at Alfred University, he had spent the final years of his career at the Brandeis University Graduate School for Advanced Studies in Social Welfare. Shortly after retiring in 1977, he and his wife moved back to the Alfred area and renovated an old farmhouse.

No one in the room had stronger credentials as a peace activist. The ethics of nonviolent protest was important to him. Although he had served aboard a small carrier in the U.S. Navy during World War II, he resigned from the naval reserves after becoming a Quaker in the mid fifties. Soon he had become involved in several peace missions including a two-year stint in Germany where he was part of a team that shuttled between East and West Germany talking to West German, French, and Russian diplomats. Recently during the Contra War he had been part of a Quaker peace mission to Nicaragua.

Roland had read about the protests against the nuclear dump in local papers. He had been intrigued by ACNAG's commitment to nonviolence, by its creativity, and by its success in keeping the siting commission from getting onto the land. A moral issue, however, had kept him from getting involved in the early actions. He couldn't shake the feeling that the protesters were simply trying to keep the dump out of their own back yards but would be happy if it could be foisted onto somebody else's.

However, after he noted that people in both CCAC and ACNAG were encouraging resistance in other counties and even other states, he began to rethink his ethical concerns. Like many others, he became convinced that the nuclear industry was trying to trick people into taking extremely dangerous material by mislabeling it as "low level" waste. Finally he decided that it was both natural and moral to defend one's own land, while fighting for principles that applied to all people. If you don't fight for your principles on your own turf, he asked himself, where *will* you fight for them? By early January Roland was active in ACNAG, participating in all the strategy sessions and standing ground at barricades in both Caneadea and West Almond.

"I think I've got a better idea. I've talked to a lot of older people who were upset that the police always passed them by when they arrested people at the roadblocks. One woman in her eighties told me she was furious when the police skipped over her to arrest someone younger. I know I could get at least fifteen older people who'd be anxious to volunteer."

I was surprised that Roland was not having any ethical concerns with wearing masks, but rather was discussing the pragmatics of the next action.

"That's great!" Jim Lucey interrupted. "The judge'll choke on this! We'll give them the respected, wise elders of the community."

"I can get old people," Roland laughed, "but I'm not so sure about 'respected' and 'wise.'" Everyone joined in the banter—genuinely for the second time in the tense meeting.

"Grandparents for the future," someone said. The name stuck.

"Judging from the mood," I summarized, "people seem to think that Roland's idea is a good one. If we decide to wear masks, we could have ten to fifteen unmasked elders openly blocking the road."

"Not that many," Stuart interrupted. "We really don't want to deplete our funds more than we have to. Three or four people seems reasonable."

"Stuart, that would be an impossible task," Roland replied. "There're too many 'grandparents' who'll want to do this. How would I choose?"

"Okay" Stuart said. "How about compromising on five or six, maximum? We'd need to have at least two, maybe three, teams at different parts of the site anyway, since we won't know for sure where they'll be coming in."

"Okay, I'll let you know in the next couple of days how many people are interested."

"Diana, does this help resolve your issue of openness?" I asked.

"It makes it better, but I still don't like the idea of wearing masks. If most people are comfortable wearing them, however, I don't want to hold up the meeting. But I'll probably choose to take a support role."

She paused. When no one spoke, she continued, "I'm still concerned that everyone wearing crazy masks could turn things into a circus."

Now she was clearly in the majority. Others echoed her concern. "Whatever we do, we shouldn't wear threatening masks."

"I don't like people choosing what kind of mask they want. We should have some standards."

Gary Lloyd continued to argue for individual choice. "I know we're all in this together, but people should be allowed to make their own decisions about what they're willing to wear. The important issue is that we can't be recognized. I wouldn't want to wear a cat mask or be a bandit like Spike. I just wouldn't be comfortable doing that. I agree we shouldn't look threatening, but let people make their own choice."

"I don't want to tell everyone what to wear," Roland Warren said slowly in his contemplative style of speaking, "but I think it would be a good thing to wear something that represented the county. For me, it was uplifting to hear arrestees saying, 'My name is Allegany County,' on the evening news. Isn't this similar? We had very pragmatic reasons for not giving the police our names, but it became a powerful symbolic statement. Wouldn't it be wonderful if we could wear masks that somehow said, 'My face is Allegany County.'"

Several people began exploring how such a mask might be created. A few began suggesting some designs.

I stopped the discussion. "Before we design the mask, we need to figure out whether it's practical to make enough of them for people to wear—we might need as many as eight hundred."

Roger Van Horn was a middle-aged Alfred resident who owned Sun Publishing Company, a small business with four employees that his father had started years earlier. His family had lived in the area for generations and he was strongly committed to ACNAG. He had coordinated the "minuteman" phone tree out of his office a few months earlier when ACNAG needed to have a place during the day where someone could always be reached. He had allowed his employee Hope Zaccagni to use time on the job to design and print posters for the anti-dump movement. He had even donated the supplies for the "County Watch" posters that now hung in grocery stores, churches, and bars. "We've got plenty of supplies. If Hope can design something in the next twenty-four hours, we can find some time at work to print them on Friday, but I'll need volunteers to come in over the weekend to cut them out and attach string to them."

"How about it, Hope?" I asked. "Do you think you can design a mask that is both nonthreatening and represents Allegany county in twenty-four hours?" She had designed and printed many of the posters, among them the bright-green County Watch poster with the image of a man holding huge binoculars that projected out from the surface.

Hope laughed, "You mean more like three or four hours don't you? If this meeting lasts like most, I won't get home until late tonight. I've got to work tomorrow—we've got a big job to finish, and I won't be able to take off any time at work. So I'll really have only a few hours after dinner tomorrow night."

"Hope, you've done great work on all our posters," someone said. "If anyone can do it, you can."

"Well, I'll try. But I won't have time to coordinate the phone tree. Someone will have to take that over for the next few days, especially if we have to get a message out to everyone."

"I'll do it," I volunteered. I was her next-door neighbor and had already helped out when the job had become overwhelming.

"It feels like we've made a lot of decisions," I said. "We'll use nonthreatening, Allegany County masks at the next action. We'll have five or six 'grandparents' unmasked, blocking the road in front of some equipment. The rest of us on the site will be wearing the masks. Does anyone want to discuss the issue of masks and grandparents any more?"

People were shaking their heads and clearly wanted to move on. Peer Bode who taught video in the School of Art and Design at Alfred University wanted to discuss ACNAG's position on videotaping the action. After getting a masters degree from the Media Center at SUNY Buffalo, he had spent several years at experimental television centers in Binghamton and Owego, coordinating programs for artists involved in electronics and video. He became concerned about the dump very early and videotaped the four-hour meeting when five thousand angry citizens shouted their objections to the siting commission at Belfast.

"Many of you probably know that some of my students have been videotaping the actions. I know everyone's worried about identifying people who're involved in the protests because of the injunction. I need to know if the group thinks it's okay for some of my students to videotape the next action."

He explained how cameras around could reduce violence by keeping the police from overreacting. "I even tell the students," he said, "that if they run out of tape they should keep aiming their cameras at the action and pretend they're still videotaping it."

In response to a question, Peer Bode explained that cameras can sometimes heighten the confrontation, because people who feel their cause has been ignored might take more risks. But that's not the case with ACNAG; everyone's committed to nonviolence and acting with dignity. If people see cameras turned on, it would just remind them to stay within the ACNAG guidelines.

"The camcorders," he said, "operate like objective observers. They're a very democratic medium." Peer was interested in the political implications of "electronic democracy," his word for the use of camcorders to further democratic principles around the world. The technology, after all, wasn't too expensive. Peer envisioned the day when computers and camcorders would be linked and people could see what was happening around the world from lots of different perspectives—including those of the "little people" who were not part of the government, news media, or police forces.

He also argued that information is power. The videotapes could be used in lots of ways. They could be fed to the news media. Someone might eventually put together a tape to share our struggle with protesters in other states. "Frankly," he said, "it's important for us to have our own information and our own records."

"The videotapes of the other actions have been invaluable to us," Sally interjected. "I've personally used them in talking to the press. I often don't even have to give them the tapes; when I say 'the videotapes show something,' that gives my comments more authority. I can't be everywhere observing an action, so I can gain a lot of information from the tapes to feed to the newspapers."

Stuart Campbell added, "We've learned a lot that helped us tactically. Remember at the last meeting we showed a tape of the arrests at West Almond that had protesters arguing with Bruce Goodale. Seeing that, we saw that verbal skirmishes could escalate. So we decided it'd be better for everyone to stay focused on the civil disobedience and remain silent."

When no one said anything, Sally concluded, "I think Peer's right. The information's really important."

"The police will be videotaping the activities anyway," someone said. "So we might as well have our own record of events."

No one in the discussion expressed any disagreement with videotaping the next action, and many expressed gratitude that Peer had organized the effort.

The meeting lasted until nearly eleven. There was more discussion about the legal implications of the injunction, about the possible use of horses to make the technical team think twice before walking onto the site, about possible scenarios for the action at Caneadea, and about backup plans if the siting commission decided to go to West Almond. No definitive action was taken on any of these issues. Since it was getting late, people decided that Spike and his strategy group at Caneadea should figure out the specifics.

We decided to activate the phone tree to call a mass meeting for Sunday. We needed to explain the legal implications of the injunction so that people could make their own decisions about participating. We also needed to explain to the wider group why it would be necessary for people to wear masks.

MARCH 29, 1990—

Hope Zaccagni was exhausted. She had gone to bed late Wednesday night after the long ACNAG meeting and had worked overtime during the day at Sun Publishing to meet a deadline. The family had eaten a quick dinner and she had put her five-year-old daughter and seven-year-old son to bed, while her husband washed the dishes and cleaned up the kitchen. She looked at her watch. It was 9 P.M.

Hope realized that she only had a couple of hours. This is going to be a "down and dirty" job, she thought to herself. Everything better fall into place pretty quickly. No time to think. No time to conceive. No time to sort out what will and won't work. Instant art!

She drew an oval that filled a sheet of letter-size paper. She made two half-dollar-sized circles at midcenter for eyes and wondered how the mask could represent Allegany County. Suddenly she realized that the eye holes could become "C" "O" and a period on the side of the "O" would turn them into an abbreviation for "County." It could also look like a tear and keep the mask from appearing threatening. It took her only a few more minutes to write "ALLEGANY" in a semicircle around the top of the mask, using letters the same size as the "C" and "O" that were traced around the eyeholes.

She paused and stared at her work. Obviously the mask needed a nose and a mouth. Looking at the stark outline, she imagined a skeleton. The teeth were next. Somehow the process seemed like carving pumpkins on Halloween. Letters would work, she thought, if they were thick enough and made a bit smaller at the corners of the mouth. "What could they say," she wondered. It would have to be short. A couple of minutes later, she thought, "NO DUMP," and began experimenting with the size and thickness of the letters until they created the contours of a mouth.

She thought of the nose as a flap that could be cut out between the eyes, and drew it in. She looked at the mask. "Too wordy," she thought to herself. She wished she could find an image that represented Allegany County. But what? It would have to be simple. The words "ALLEGANY CO." worked, but an image could replace the "NO DUMP" teeth or become incorporated into the nose and forehead. Hope had been working more than an hour. She felt a bit chilled in the cool house and went into the kitchen to make herself a cup of herbal tea.

Returning to the work table she had set up in her living room, she looked at the mask. "Eureka!" An image appeared to her. The nose and the forehead seemed to form a mushroom! Perfect! Protesters in the county had jokingly referred to themselves as mushrooms ever since Angelo Orazio made the statement to the press that their leaders had been treating them like mushrooms. She also liked the double meaning and thought of the mushroom clouds of nuclear explosions. Even better, the image was simple and could easily stand out on the face, if it were printed in a bright color. She took an orange marker and filled in the mushroom space on the white sheet of paper.

Then doubts set in. The mask seemed too simplistic, too plain. She needed to try another design. She looked at her watch. It was 10:30 P.M. She walked into her kitchen and looked out the window toward her neighbor's house. The lights were still on. She picked up the phone and called me. "Can I come over and show you a preliminary design?"

A minute later Hope was standing in my dining room showing me the drawing. "I don't know yet what colors we'll use, but I want the mushroom to stand out. It'll depend somewhat on what we've got in stock at Sun Publishing, since there's no time to order anything. I'll keep the lettering black against a light background color and the mushroom design a brighter color. I don't like the white—it's too ghostly. I'll get a warmer color, maybe fleshtone."

She waited for my response.

"I'm amazed. You've done it! There's nothing threatening about this mask. It even looks a little sad with the tear." I laughed, "It's goofy enough that people in the county will relate it to Orazio's mushroom comment. Given the instructions of the steering committee, I think it's brilliant." I was not just massaging her ego. Knowing how much pressure she'd been under, I was impressed that she'd been able to come up with so successful a design on such short notice.

APRIL 1, 1990—

More than half of Holmes Auditorium on the Alfred University campus was filled on Sunday evening—at least 350 people. I had never before tried to moderate such a large meeting. The size, however, was less significant than the difficulty of negotiating the issues. I needed to be absolutely clear about the legal implications of the injunction. I also had to explain why the steering committee had decided it was necessary to wear masks without divulging the specific details of the action.

I quickly explained how the injunction had changed the consequences of our civil disobedience. There were many questions. I could easily explain the one thousand dollar fine and six months in jail, but was unsure how the judge might interpret "damages" that the state would incur because of our actions. I presented the purported information from our "source" that Judge Gorski wanted to finger fifty protesters so he could levy the maximum fine and break the movement. I explained why the steering committee had reluctantly de-cided that ACNAG would have to use masks in the next action.

A couple of people challenged the decision of the steering committee with arguments similar to those used by Diana Zweygardt. "I certainly don't want to impose arbitrary decisions on the group," I said, "but we discussed the issue thoroughly at the steering committee meeting a couple of days ago. No one there really *liked* the idea of wearing masks, but when push came to shove almost everybody believed that we had to do it. This was not a small meeting of a few people—more than fifty people were there.

"One of the reasons for the meeting tonight is to let you all know what's going to happen so you can make your own decisions—both about the legal liabilities and the masks. Civil disobedience is always a personal choice, and because of the new situation, you may decide to change your role from being on the front lines to being support. At some point, people have to be on the same sheet of music, however.

"There's probably someone here who's a police informer, so I can't really get into all the details.[2] I can only say that we plan to have designated people in the front lines without masks who'll be arrested. We hope that'll preserve the principle of openness without playing into the attorney general's trap. If you can be identified through photographs, it's the same thing as being arrested—in terms of the injunction. So please don't get involved in this action if you've got moral scruples about wearing masks. We're going to have enough difficulties in this action without being divided."

At the end of the meeting people saw Hope's mushroom mask for the first time. There were black letters on a soft, yellow background with a bright red mushroom in the center, defining the nose and forehead. "This will visually portray what we told the police when we were arrested—'My Name is Allegany County.'"

❖

APRIL 4, 1990—

The evening before the siting commission came to the county in full force, the phone rang in the Campbells' house. Helen Hutchinson, coordinator of the ACNAG phone tree in the Wellsville area, was on the line. She asked Stuart whether he and Sally were still looking for horses. She told him that Franny Root had eight prize Belgian horses, worth thirty thousand dollars each, that he was willing to bring to the protest—along with riders. Although Stuart didn't know it at the time, Root's large horses were famous for winning prizes in weight-pulling contests at the county fair. Root was a well-puller. Occasionally to unclog an oil or gas well, the casings need to be pulled out; well-pullers today mostly use bulldozers, but Root had stuck with his horses. Had he been an artist or a poet, he might have perceived the horse as a metaphor for greatness and power.

On many occasions Root had come into conflict with the oil and gas companies. A number of times he'd stood up to them when some bureaucrat tried to manipulate him. He loved a good fight, especially if it was against people who wore suits to work. As a rugged individualist, he usually fought

---

2. As incredible as it seems, subsequent events seem to show that both the attorney general's office and the state police were completely taken by surprise that protesters would be wearing masks at the action. The sheriff, of course, knew that we might wear masks, but asserts that he had not heard any more of the matter.

his battles singlehandedly, but when he heard that ACNAG was looking for horses, he was willing to lend them.

"I'll get back to you, Helen, after I talk with Sally and Spike," Stuart said.

Spike had raised the possibility of using horses at the end of ACNAG's last steering committee meeting. Over the next few days Sally had become obsessed with getting horses, even as Spike turned his attention to other matters.

I had driven with Sally to Spike's house in Belfast the day after the ACNAG steering committee meeting to explain Timberlake's concern that the police might just walk around the roadblocks and escort the technical team to the site. Strategizing, however, was never one of my strengths, and Spike paid even less attention to me than he did to Sally. I only halfheartedly pressed the case anyway, because I felt we had only two alternatives. Either the roadblocks would work, or we'd have to encircle the siting commission.

"Trust me, Sally," Spike had said. "I've planned for every contingency. It doesn't matter whether they ride or walk. They'll still have to come to a series of roadblocks and read the injunction at each one. We'll have enough roadblocks that it'll take them all day to get onto the site."

That night at home, Sally began to press Stuart to get horses. Like Spike, he thought it unlikely that the police had changed the rules and expected the troopers to operate as they had in the past.

"It doesn't make any difference whether Timberlake had any inside information," Sally said. "He's right. They're going to march onto the site! Maybe if they see horses, they'll think twice. Timberlake's told us what's going to happen. The cops aren't going to play by the rules this time. That's what Timberlake was trying to tell us! We've got to get horses! That's the only thing that'll save our ass. Why aren't any of you guys listening?"

"C'mon Sally, that's not fair," Stuart replied. "Because of your arguments, Spike's pushed the first roadblock back to the Caneadea bridge, three or four miles from the site. Don't tell me the troopers are going to abandon their cars and walk that far. It'd be nuts!"

"Then Gary's right," she insisted. "They'll come in the back door, and there's no way we can push the roadblocks any further out on the west side of the site—there are too damn many intersections. If we're successful in keeping them from using the injunction, they're going to walk."

When no one seemed to listen, Sally became obsessed with getting horses. Stuart knew that when his wife became determined, there was nothing he could do but go along for the ride. He, too, asked people about getting horses.

The day before Helen Hutchinson's call, however, he had promised Kay Eicher at a Caneadea strategy meeting that he wouldn't use horses at the roadblock itself. "Stuart," Kay had said, "horses are just too dangerous. People could get hurt."

The promise was easy for Stuart to make. He was sure he wouldn't be in charge of anything, planning, as he was, to be back at the obscure corner of the site where he'd been the last time. Spike, he knew, would be calling the shots. And since they'd only found a handful of horses anyway, the issue seemed academic, and he remained convinced the troopers would not walk around the barricades.

Now, remembering his conversation with Kay, he was a bit uneasy as he went upstairs and told Sally about Root's offer. They called Spike, but Sally only stayed on the extension for a short time as her anxiety rose with the two men talking as though there would be no surprises.

"It can't hurt to have more horses," Spike told Stuart, concluding the conversation. "Have people bring them to West Hill and Klein."

Stuart returned Helen Hutchinson's call and told her to accept Root's offer. "You were at the steering committee meeting. Explain the necessity for wearing masks and remind him we're committed to nonviolence."

Sally slipped out the door to check the weather as she did every night before an action. A cold northerly wind was blowing and it felt like snow, but she shivered more from the dread of tomorrow's menacing confrontation than she did from the weather.

9 Grandparents, Horses, and Masks

In that early morning gloom I felt very uneasy surrounded by yellow masks concealing the silent urgent faces I've gotten to know so well over the year. I can't quite explain it, but the feeling haunted me all day, whenever I looked at those expressionless, passionless, yellow orbs."

—Kathryn Ross, of the *Wellsville Daily Reporter* in column "My Home Town" on April 9, 1990

APRIL 5, 1990—

SALLY CAMPBELL WAS CLEARLY WORRIED as she walked toward the Old German Church, which was the protesters' main staging area for the looming confrontation. She heard the drumming of a police helicopter and watched it circle the area for two or three minutes before it flew off to investigate another part of the site. A half-hour earlier she had driven by a huge mobile command center, filled with sophisticated communication equipment that the state police had last used during their fierce confrontation with the Mohawks at Akwesasne.

A half-foot of wet snow had fallen during the night and gusts of wind were blowing snow in Sally's face as she confronted a bleak world devoid of streetlights. Car lights pierced the inky darkness as protesters arrived at the tiny white church that was a monument to the religious faith of the farmers who had settled the area. Monitors were handing out yellow-orbed mushroom masks to everyone who entered the area, explaining again how the injunction had changed the rules of the game and why they should wear masks to keep from being identified.

The night before, ACNAG leaders had preassigned a couple of hundred of their best-trained people to go to specific places around the site. One hundred of them were already at the narrow bridge that spanned the Genesee River at the Village of Caneadea. The sheriff had told the protesters that the siting commission would not fool around going from roadblock to roadblock this time, but would make a direct assault on the site across the bridge, arresting any protesters who were in the way.

Sally thought that Spike's plan of barricading the road with heavy farm equipment would work if the police stopped to remove each barricade before going on to the next. But she doubted this would happen. Ever since Timberlake had suggested that the troopers would "march onto the site," Sally feared that ACNAG's strategy wouldn't work. But she couldn't counter Spike's logic that it would be insane for the police to abandon their vehicles and support units three miles away from the site. Videotapes of past actions clearly showed that the state police hated being flanked in the rear by protesters. Grudgingly agreeing that Spike was probably right, Sally then decided the troopers would attack the eastern side of the site where roadblocks were only a half-mile away. She hoped, without much confidence, that the two dozen horses would convince the state police to stick with their original plans. The poor roads and freak spring snowstorm might further encourage them to use the direct route across the Caneadea bridge.

As logical as these arguments seemed, Sally feared there would be violence; something unpredictable was going to happen. Reliable sources, she'd been told, reported that there were nearly a hundred state troopers from all around the state. The troopers were clearly making an intense effort to get the siting commission on the land this time.

Sally watched the crowd at the church grow from one hundred to several hundred people in less than half an hour as the gray light of dawn slowly dispelled the night. ACNAG's success in attracting crowds was also its Achilles heel. Sally wondered how many of these people had been to an ACNAG training session. She hoped ACNAG's strategy of keeping most of the folks up by the site at the old church would work. Drew Robinson, one of ACNAG's most burly guys, was monitor there. He was initially joined by Stuart Campbell whose power of persuasion was indirectly proportional to his diminutive size.

Sally walked up to Drew who was standing beside his truck. "Keep everyone here informed about what's going on," she said. "Then people won't be as tempted to drive all over the site or decide to go down to the bridge." Sally then got into her truck and drove to check out other parts of the site. As one of the specifically named individuals in the injunction, she knew she wasn't supposed to go down to the bridge, but she decided to go anyway. As media coordinator she felt she needed to know what happened at the confrontation.

As she was leaving, she watched Stuart, wearing a mask, pick up the bull-horn and address the crowd. "We're pretty sure that the siting commission will be coming across the bridge at Caneadea. That's the most direct route to the site. The snowstorm makes it less likely that they'll come in the back door. Raise your hands if you're willing to be arrested." Perhaps two hundred people raised their hands. "We want to minimize arrests today. Judge Gorski wants judicial hostages and we're going to give him some, but not ones he wants. We have a few elderly people who've chained themselves to the bridge. The state police won't want to deal with grandparents, and I don't think the judge'll want to deal with them either."

"We'll stop them if they get up here to the site," someone shouted. Several others affirmed their determination by raising their fists into the air and shouting their defiance.

"We've got about one hundred people down at the bridge," Stuart continued, "and a few more at various points along East Hill Road. We've got tractors and farm equipment, blocking the road at several places. There'll be one roadblock after another. After the injunction's read, people will retreat. It'll take the police all day to remove the equipment from the road. Some of the stuff is so heavy that ordinary tow trucks won't be able to move it." Always the European historian, Stuart added, "It'll be like the Russian retreat in World War II—a long series of barricades.

"You're needed here as the final line of defense. Should the police get by our roadblocks, we'll let you know where to go to make our final stand."

Stuart hoped that it wouldn't be necessary to use the people here. But Sally's anxiety was contagious. What if he and Spike had too cavalierly assumed that these people wouldn't be needed? Shouldn't they have made firmer plans? Nervously he put down the megaphone and walked over to Drew, who was listening to the CB radio in his truck. "Everyone seems to be in a good mood, but the masks make it hard to tell who we're dealing with."

"I know quite a few people here," Drew said. "They're okay. I'm more worried about the ones tearing around the site. I saw a car a few minutes ago, filled with people drinking beer and whooping it up. And it was only 8 A.M.!" Just then a truck roared by, filled with people in Halloween costumes, shouting and waving at people in the churchyard. The driver leaned on his horn and sped down the road.

"I wonder how many people like them have already gone down to the bridge?" Stuart worried aloud. "Let's just hope the rowdies don't start something."

"I've been talking to people here. Quite a few aren't ACNAG regulars; some aren't even from the county. But they don't seem violent."

Stuart knew that there wasn't anyone at East Hill and Pinkerton. That was the intersection at the edge of the plateau about one-half mile from the church where police had arrested people in January. It was also the gateway to the site from the Caneadea bridge. (Map of the Canadea site, p. 152.)

Stuart said to Drew, "If you think you can handle this crowd, I'm going to go try to stop more people from going down the hill to the bridge. I'll keep in touch with you on the CB."

Stuart heard the helicopter buzzing over the churchyard as he got into his truck and drove west along Shongo Valley Road a quarter-mile to the corner of Pinkerton. At least the police will know there are hundreds of people up at the site, he thought. As he turned left, he noted that there were at least fifty or sixty people, standing around a bonfire. He wondered if there was a monitor and whether these people had attended an ACNAG training session. He didn't stop, but continued down Pinkerton another quarter mile to East Hill Road.

No protesters were anywhere in sight. There was only a car parked on the other side of the intersection. "I'll be damned," he said to himself, "if it isn't Alton Sylor!" Although Stuart was a rational man who didn't give much credence to the supernatural, he wondered whether this was an omen. The last time he had seen Sylor at an action was when the conservative county legislator ran into the Millers' house to inform the protesters that the technical team had gotten onto the land. That action had proven the effectiveness of civil disobedience to people in Allegany County. Now Stuart wondered whether this would be the definitive battle between ACNAG and the siting commission.

Snow flurries continued and the temperature hovered below freezing as the first protesters arrived at the single-lane, steel-deck bridge. A gale-force wind whipped along the Genesee River and up through the bridge's rusted deck. The tiny village of Caneadea spread along the west side of the river. On the east side the road climbed for nearly three miles through farmland and forest to a broad plateau where the site was located.

An ACNAG "stage crew," wearing masks, was preparing for the confrontational drama between police and protesters. Six elderly men and women nervously waited in their heated vehicles for the beginning of the first act of a performance that might destroy them financially. Two men lugged pieces of plywood a few yards onto the bridge and wired them to the deck so that the grandparents' folding chairs would not get caught in the grating. A farmer scaled the girders and tied an enormous American flag above the planking across the top of the bridge. Farmers with meager resources momentarily ignored legal risks and supplied costly farm equipment for roadblocks. Protesters' nightmares of losing their homes temporarily receded into the background as they prepared to defy the siting commission once again.

When tailgater announced that the siting commission had left the sheriff's office, tensions mounted. The crew hurriedly set chairs on the planks and stretched a heavy chain across the span, padlocking it to girders on both sides. Protesters escorted the grandparents from their cars toward the bridge, as though they were at an odd masked ball where everyone wore the same costume.

A small, perky eighty-seven-year-old woman, dressed in a black wool coat and warm matching hat, was the first to arrive and take her seat. Though a small woman, Alexandra Landis radiated determination and strength. An independently minded woman, she frequently was the only opposition vote when she served on the Alfred-Almond school board. Years before, after receiving her doctorate from Harvard University, she had taught philosophy, aesthetics, and art history at Harvard, the University of Miami, and Syracuse University, where she had been chair of the art school.

Clarence Klingensmith, a sprightly seventy-four-year-old emeritus professor of chemistry at Alfred University, quickly followed her and took a seat. Neatly dressed in a dark overcoat and gray wool cap, he had a tie poking from beneath a wool scarf. A naturalist all his life, he had a government permit to band birds. In retirement he continued to research the buffering effect of clay soil on the acid rain falling in Allegany County from coal-burning plants in the Midwest.

Both Alex (as her friends called her) and Clarence regularly attended the Quaker meeting in Alfred, which Roland Warren had helped organize in the 1950s. Roland now took a seat in the middle of the bridge between his two friends. The same age as Clarence, he was a tall, robust man with thick gray hair and a determined chin.

Roland chuckled when he recalled asking Alex whether she wanted to participate. "Of course," she answered. "I was standing in the front line at West Almond, but the police wouldn't arrest me. Couldn't I sue them for age discrimination?"

"Listen, Alex, this time it's really serious. They're going to use the injunction. That means we'll probably go to jail for thirty days."

"I can do that." She paused, then asked, "Do you think they'll let me have a copy of Shakespeare's works? I can put up with anything then!"

"I don't know, Alex." Roland laughed. "I can't imagine anyone daring to refuse such a request, coming from you! But it's not just jail. The judge may give us heavy fines for damages. That could mean thousands of dollars."

"As long as they allow me to read, I'm willing to chance the money. The siting commission is taking away our freedom. What's the point of having money and living in a dictatorship?"

Roland had gotten similar, though less colorful, responses from everyone he talked to. He formed two teams of elders, one for the bridge, the other at the western edge of the site.

Ermina Barber was pushed in her wheelchair to the south edge of the bridge. Though plagued by a serious case of arthritis, she wanted to take a stand "for her grandchildren and future generations." She had lived in Allegany County all her life, and most of her family had settled here. She carried a huge baby picture of her daughter who was now due to deliver Ermina's fourteenth grandchild.

The wind blew directly on Ermina. A mushroom-masked woman stepped between her and the side of the bridge to absorb some of the chilling gusts. Wind filled the American flag as though it were a sail above their heads and noisily whipped through the grandparents' clothing. The Genesee River swirled below.

William Parry was a well-known ceramic sculptor whose work frequently referenced people's relationship to their environment. At age seventy-one and retired from the School of Art and Design at Alfred University, he sported a neatly trimmed white beard and wore a brown leather overcoat and cap. A soft-spoken, congenial man with a broad smile, his eyes twinkled as he answered reporters' questions and awaited arrest.

Henry Koziel, a sixty-four-year-old farmer, owned property on the site. That very morning he had insisted on participating, apologizing for being the "juvenile" of the group. "I'm nearly sixty-five. I live right here on the site. So you'll just have to accept me!" He refused a chair, preferring to stand by the side of the bridge. But he took a pair of handcuffs and linked himself to the chain, as had the other five.

I had been designated as monitor at the bridge over the objections of Spike, who as site coordinator wanted to be there, injunction or no injunction. Spike had made arrangements for an ACNAG spokesperson to meet with the sheriff when he arrived on the morning of April 5. The sheriff, eager to discover the protesters' plans, had readily agreed that this person would not be arrested and could wear a mask.

Just two days before the encounter, Spike had tried to convince Stuart, Sally, Gary, and me that *he* could get the sheriff and police to back away from the confrontation, once they realized what they were up against.

"Spike, you're certainly the most charismatic and persuasive person in ACNAG," said Stuart. "But the state isn't spending all this money to turn around at the first roadblock because some silver-tongued devil says something convincing."

"Maybe. But what would it hurt? Larry's agreed not to arrest me. I'd be wearing a mask. The police don't want to be there. If they're convinced the task's impossible, they might back off."

"Since the siting commission has pegged the four of us as ringleaders," Sally argued, "none of us should be anywhere near the roadblock at the bridge. Judge Gorski would like nothing better than to get one of *us*. That would make his day."

"I'd just speak with the sheriff and then leave."

"I don't get why you're so invested in this, Spike?" Stuart said. "Tom can tell the sheriff what's happening. Larry would like nothing better than to leave, but you know it's not in his hands. The siting commission and the attorney general are calling the shots."

"Look, Spike," Sally said. "We told everyone who was specifically named in the injunction to stay away from the action if possible. And you've got to abide by the same rules."

"Besides," Stuart added, "you're the only one who knows all the plans. You can't risk getting caught, especially when it's so unnecessary."

Spike reluctantly gave in.

Shortly after 10 A.M., the sheriff pulled his car over to the side of the road in the village of Caneadea, within half a block of the bridge. Wearing the pale-yellow mushroom mask and a bright yellow hat and scarf that I'd bought in a secondhand clothing shop, I took a deep breath and walked out to meet him. The sheriff opened the back door and gestured for me to get in.

"I'm not speaking to any masked man!" objected a round-faced, short man sitting in the back seat. "If he wants to talk to us, he'll have to take off the mask."

"If you don't want to talk to me, I'll just turn around and go back," I responded hotly. "I thought you wanted to know what's going on."

"Please, let's all calm down," the sheriff interjected. "This is Mister Peter Eiss who's with me today as an observer from the attorney general's office."

Then he turned toward the man who was seated in the car and said in a soothing, yet determined tone, "Mister Eiss, please! I want to hear what this person has to say. I asked the citizens to send someone to talk with me. It's important we have all the information we can get."

Since reporters hovered around the car, the sheriff again gestured for me to climb in the back seat. I sat down, trying to ignore Eiss who glared coldly at me.

"There are six elderly people sitting in chairs, blocking the far side of the bridge," I said, beginning my rehearsed monologue. "Most are in their seventies;

one woman is eighty-seven years old. Another woman is in a wheelchair. Behind the grandparents, farm equipment's blocking the road at the edge of the bridge. We've got fifty or sixty people, wearing masks, behind the equipment. They're there to support the grandparents, but will leave when arrests begin." Having successfully delivered my first lines, I felt some of the tension leave me, much as it always did after a first scene in a play. I took a deep breath and settled more comfortably into the seat.

"That's only the tip of the iceberg," I continued. "You'll encounter lots of roadblocks. The police helicopter's flying around the area, so you probably know that there are hundreds of people up at the site itself and lots more along the road."

"Actually I haven't been listening to the police frequency, so I don't know what the helicopter's reporting," the sheriff said. "What can you tell me about the other roadblocks?"

"There'll be lots of tractors and farm equipment. Masked protesters will be at each. They'll listen to the injunction and then move up the road, leaving the equipment behind. If you get close to the site, there are more grandparents who're willing to be arrested. You should go back and talk all this over with the police, since there's no way the siting commission's going to get on the site today."

"Are the elderly citizens okay?" the sheriff asked.

"They're warm enough. None of them have ever been arrested before so they're a bit nervous, but determined."

"Is there anything else you want to tell us?"

"Not really. But there's no way you're going to get close to the site! So why don't you turn around and forget it?"

"That's up to the siting commission and the police, but I'll tell them what's going on."

Walking back to the bridge, I saw a large banner unfurled behind the elders with foot-high letters, "GRANDPARENTS FOR THE FUTURE," and the huge American flag billowing high above their heads. "What a photo that'll make," I thought.

Getting closer I heard a reporter talking to Alex Landis. "How do you feel about all this?"

"Well it's just one of those things you have to do."

"Are you afraid?"

Raising her shoulders and straightening herself in the chair, she shot back, "No! Not a bit! I don't get afraid of people who do stupid things."

Another reporter was interviewing Ermina Barber. "You could go to jail for this."

"Oh, I know that," she said, shrugging her shoulders.

"Do you want to go to jail?"

"I don't want to go to jail, but I'm willing to go to jail to get my point across."

"What's your point?"

"No dump in Allegany County or any place else in the nation as far as I'm concerned. I'd like to have them store the stuff where it's at." She paused. "We've got to change all our ways of living so that we can get back to the way it was. I remember Allegany County when it was so clean, so beautiful. We've got to change our lifestyle." She continued, struggling to hold back tears, "I have arthritis very bad and shouldn't be out in the cold, but I'm willing to lay down my life for Allegany County and my grandchildren."

Bill Parry was talking to another reporter. "This is a political solution to a scientific problem that right now has no scientific solution."

With cameras whirring, Roland Warren admitted that the blockade of grandparents was his idea.

"Who got you to come do this?" a reporter asked.

"Well, as a matter of fact, it was my suggestion that we do this. Then everyone volunteered."

"Have you ever broken the law before?"

Always thoughtful, he smiled and replied, "That's a hard question. I've never broken the law deliberately in a demonstration before. I've never been arrested."

Just then, the sheriff arrived. He looked at the six elderly people sitting in the middle of the bridge. "How are you? Are you staying warm enough?"

Roland answered for the group. "We're all comfortable enough, I think. Thank you."

"I have Mister Peter Eiss with me from the Department of Law. We wanted to walk out and see how you are. Is everyone comfortable?"

The elders nodded and one of them said, "We're all right."

Scholes and Eiss conferred together briefly. Eiss turned to the elderly protesters, "Are you going to move?"

"No, we're not going to move," Roland calmly replied. "In fact, we can't move. We're handcuffed to this chain. We're willing to be arrested if it comes to that. There are a few hundred people behind us and a few thousand more in Allegany County who'll take their place any time it's needed."

The sheriff turned to the lawyer from the attorney general's office. "I assume we need to go back and talk." Eiss nodded, and the sheriff turned back to the protesters. "We're going back to assess the situation. You'll be warm enough here?"

Roland looked up and down at everyone in the group, and responded. "We're comfortable enough. If we could trust you folks, we might take shelter, but we'd rather stay here than let you on the site."

Fifteen minutes later the sheriff returned and asked the media to step back. He and a couple of his deputies then began the scripted ritual. "The siting commission's in a vehicle just down the road, and they're insisting I enforce the court mandate and put them on the land. I'm asking you now to please open up the road and allow the technical team on the site."

"We can't let you in," Roland replied. "You're doing your job and we respect you for it. But we're doing ours and we're not going to let you in."

"Okay. Then we need to read you a statement that was prepared by the attorney general's office, at which time I'll serve each of you with a copy of the injunction."

Scholes's deputy took less than two minutes to read the statement: "You have the right to protest as long as you don't interfere with the work of the siting commission. . . ." The judge may give you a maximum thirty days in jail and fine you one thousand dollars and an additional amount for damages. "This could require you to pay for the siting commission's extra costs caused by your actions. . . ."

"Do each of you understand what was read?" the sheriff asked.

"I didn't hear it very well," Alex said, in order to slow down the process further. "You'll have to read it again."

A bit surprised, the sheriff asked, "Would you like it repeated?"

"Yes, please."

"Very well, ma'am."

After the second reading, the sheriff asked, "Will you now open the road for the technical team?"

"No. We're not prepared to move," Roland responded.

"Then I must serve each of you with a copy of the preliminary injunction."

"Could I give you something first?" Alex asked.

"Yes, ma'am."

She began speaking, while simultaneously struggling to get a bulky object out of a large paper shopping bag. "I want you to know that I consider my freedom is being taken away from me, and I want to give you something. You can give it to Mister Cuomo and he can give it to Mister Bush. This was given to me. I think you'll understand what it means." She pulled out a large, folded American flag that had been on the casket of her son Robert, who had been a navy pilot in World War II. "If our freedom is taken away from me like this, I don't want this thing any more. Help yourself."

The sheriff, visibly shaken, put the flag under his arm and gave everyone a copy of the injunction. "It's a cold day. I don't like to see you sitting out here like this. Please let's remove the chains and get you people in where it's warm, and let us continue on to whatever the next obstacle might be."

"We can't do that," Roland responded again. "I'm sorry we've got such rotten weather for you, but that's the way it worked out."

"The weather doesn't bother me," laughed the sheriff.

The sun came out for the first time that day. Almost immediately, water from thawing ice began dripping on everyone.

The sheriff turned again to Alex. "Ma'am, I'll see to it that the state authorities get this." Then he left to get the police.

Reporters asked Alex questions about her son and her motivation for relinquishing the flag. "How does it make you feel to get rid of the flag under these circumstances?"

"It doesn't stand for anything anymore. There's a quotation from Shakespeare, *Richard III,* I think, that says, 'He daubs his vice with the pretense of virtue.' That's what's being done with the flag right now! It's a symbol of liberty, you know, but we're having our liberty taken away from us."

At 11:30 A.M. the sky darkened and snow again started to fall as the state police marched up. Lieutenant McCole approached the elders and spoke directly to Roland. "Perhaps I'm jumping to conclusions. Am I correct in assuming that you've decided not to leave?"

Roland nodded his head.

McCole gestured for one of the troopers to come forward. He was carrying huge bolt cutters that were nearly as long as he was tall. The trooper easily snipped through the heavy chain. "Officer," Alex quipped, "I hope you can cut the psychological chain too."

One by one McCole asked the elders their names. Each answered, "My name is Allegany County," and police officers escorted the elderly protesters to the police van. Twice, McCole insisted that Alex get into a wheel chair. Each time she vigorously refused, shaking off the arm of the arresting officer. "I'm *not* going in a wheel chair! I'm *walking!*"

The last grandparent was arrested shortly before noon, and the masked supporters retreated to the second barricade.

Spike Jones had been organizing farmers for the previous five days. When he fell exhausted into bed the night before the confrontation, he thought everything was settled. He knew where the roadblocks would be and where each farmer would be bringing heavy equipment. This morning, however, when he checked with them again, he discovered that a couple had changed their minds and would "rather not" commit some of their most expensive equipment. Frantically, he spent the early morning hours changing plans and finding other equipment for roadblocks. By the time the sheriff arrived at the bridge, things finally seemed settled.

When the police committed themselves to coming up East Hill Road, Spike had renewed confidence in his plans. He could forget about using the horses on the eastern part of the site; he wasn't entirely sure how he'd use them anyway. Now he concentrated on his plan to put the next barrier in place, as police removed the previous one.[1] He planned four barriers on the first half of East Hill road—a heavy field tiller at the bridge, hay baling machinery at the second, a large tractor and hay wagon at the third, and a manure spreader, filled with pig dung, at the fourth. Spike was fairly certain that the police wouldn't be able to remove the equipment from the four roadblocks before nightfall, but if they moved faster than expected, he could call in equipment from other places.

Spike was sitting in his old van near the third roadblock when the radio reported arrests had begun at the bridge and the protesters were retreating. He called Ange, the stocky monitor who had coordinated ACNAG's communication system at Caneadea, and told him to move in equipment at the second roadblock.

Nearly an hour passed before he heard that the sheriff was reading the injunction at the second roadblock. He smiled to himself. The police had taken even longer than expected to remove the field tiller that was blocking the bridge; the heavy hay binder and large tractor should provide an even greater challenge.

Spike saw me walking up the road. Puzzled, he got out of his van. "What the hell are you doing here? You're supposed to be with the people at the roadblock."

"I thought Ange was the monitor there," I replied. "I heard there were lots of people along the road and thought I should make sure they all knew what's going on."

"No, Ange is our radio contact there. He's supposed to let us know what's happening. You were to stay with the group and tell them when to retreat!"

I hopped into Spike's van. He called Ange on the CB. "What's going on down there?"

"The sheriff just read the injunction for a second time. The helicopter kept buzzing the area and protesters said they couldn't hear it. Everything's peaceful. People are standing at the roadblock."

---

1. Spike did not want to put permanent roadblocks along the entire route of East Hill Road in the early morning, because he didn't want the police to change their minds and come into the site at a less-guarded entrance. The police helicopter was reporting numbers of people and types of barricades that were in place around the site. So Spike hid the equipment in driveways or left it sitting in fields near the road. It would take no longer than five to ten minutes to get each of the barricades in place.

"Tell them to retreat when the sheriff asks them to open up the road, then call me back."

"Roger."

"What's happening between here and there?" Spike asked me.

"There are lots of people. They must not have paid any attention to directions about staying up at the church. Just past the second barricade, a group of fifteen or twenty are making huge snow boulders and rolling them onto the road. They were all wearing mushroom masks. I told them to be sure to stay out of the cops' way. I tried to get people milling around in the road to go up to the church, but they didn't pay any attention to me. The masks make it weird. I don't have any idea who I'm talking to, and they don't know who I am. I didn't take off my mask, because cameras are everywhere. Just past the snow boulders was a woman and her kid—maybe four or five years old. They were sitting on a rock, eating a lunch."

"Great time for a picnic!" Spike remarked.

"The woman looked at me as though I were from Mars when I told her she should take her kid back up to the church. I took off my mask and told her she'd better leave soon or she wouldn't be able to move her car. I think that got her attention."

Just then the radio crackled. "Troubleshooter to Sarge."

"Yeah, I'm here. What's going on down there?"

"The sheriff asked everyone to clear the road. People said they weren't moving, so he went off to get the troopers. I told them to move, but a woman who wasn't wearing a mask said, 'No! We're staying right here.'"

"Okay. Stay on the CB. We'll be right down."

"Shit! I'm sorry Spike. I should've been there. You're not supposed to get that close to the action."

"I'll wear the mask and park the truck heading uphill for a getaway. Let's just hope we get there before the state troopers. If they start arresting people who're blocking the road, all this mask shit is for nothing."

Spike parked his van near the snow boulders, about one hundred yards up from the second roadblock. Slamming the transmission into reverse, he jumped out of the van and ran downhill, his green plastic poncho flapping behind him. No police were in sight. "People!" he shouted, "We're not looking for arrests right here. Please move back." Some recognized his voice and began walking up the hill. Others, however, remained behind the hay baler and seemed determined to be arrested. Spike ran up and said, "People, please move back to the next barricade." Two people from outside the county weren't wearing masks. Spike recognized a leader from Cortland County, where they were also fighting the siting commission. Taking a chance, he lifted his mask, looked the fellow in the eye, and said, "Listen!

Snow Boulders: Using Natural Resources © Steve Myers 1990; all rights reserved.

We don't want arrests here. Please get the hell off the road or follow the others to the next roadblock!"

Just then Spike saw troopers approaching in the distance, and he shouted one last time to stragglers, "Please get off the road. Move up to the next barricade as quickly as you can, or at least get the hell off the road!"

Fifteen minutes earlier, when the protesters told the sheriff they wouldn't move, he and his deputies retreated several yards downhill to await the troopers. The midday sun had warmed things considerably, and Scholes unzipped his fleece-lined leather jacket. A small group of reporters and a few protesters surrounded him to find out what was happening.

A bystander asked him whether he would run for the new county executive position that legislators were considering. "You'd be a strong candidate."

The sheriff laughed, "I'm just trying to win reelection for sheriff this year."

"No problem! No problem!" shouted a woman in the crowd.

A newspaper reporter asked, "Will you run this year?"

"Yes. I'm up for reelection."

"Tough year for an election!" someone quipped.

Scholes smiled. "Is that timing? or what!"

The good-natured banter ceased abruptly as a column of state police came into view. The sheriff, seeing McCole, walked downhill to meet him. "At this point it looks like we're going to have to make arrests," the sheriff said.

"We're going the rest of the way," the lieutenant said.

"What?" Scholes asked, puzzled.

McCole averted his eyes, mumbled, "Yeah," and started walking uphill.

Scholes had worked with McCole for more than fifteen years and had never seen him looking so stunned. The sheriff knew the troopers were trying to get the technical team onto the site. So what was McCole telling him? Something, he knew, had changed. Just then he saw Sullivan, the head of the attorney general's office in Buffalo. What the hell was he doing here? "Something stinks," the sheriff thought to himself.

The sheriff followed the state police to the roadblock. The troopers didn't hesitate a second, but marched around the farm equipment in the road. The sheriff suddenly understood what was happening. The police were going to walk the three miles to the site, without first removing the roadblocks. Scholes couldn't believe anyone would make such a dumb decision. Ordering the troopers to hike that far without any support vehicles seemed like madness, he thought. What if the protests grew ugly as they neared the site? An ambulance couldn't even get there.

Scholes wondered who had ordered such a dangerous course of action. Certainly McCole wouldn't have ordered anything so foolish, he thought.

The sheriff suddenly understood why the lieutenant seemed so upset. His bosses in the huge communication trailer, parked in a gravel pit near the village of Caneadea, must have ordered him to do this. The sheriff involuntarily shuddered; he had a strong premonition that someone would get hurt before the day was over.[2]

Spike grabbed my shoulder when the troopers marched around the hay baler without hesitating. "Let's get the hell out of here!" Spike cried. We raced up through the fields and jumped in the van. I turned around and saw the police walking around the snow boulders as Spike stepped on the gas.

The sun had turned the plowed dirt road into squishy, springtime mud, and the van slid from side to side as it lurched up the road. Spike parked up-hill from the third barricade and told a farmer to pull the large hay wagon across the road. Just after two young men unhitched the wagon and climbed back into the cab, the troopers arrived. "Move this out of the way," one trooper shouted. McCole arrived a few seconds later and told the two men they were under arrest. When they hesitated, he told them to climb down immediately or they would not only be charged with disorderly conduct, but also with resisting arrest.

Although the protesters hadn't had time to form a human barrier at the hay wagon, the troopers began arresting anyone close by. Throughout the day, Sue Beckhorn, now co-chair of CCAC, had known that she shouldn't get arrested, so she carefully stayed off the road whenever the police were near. Wearing her tricorn hat, she now stood with her guitar in a field beside the road, singing her Allegany County protest song. A trooper approached, tapped her on the shoulder, and arrested her. Glaring at him without saying a word, she trudged down the hill. Fleurette Pelletier, a woman in her late sixties, had been singing beside her. She argued with the trooper, "Since when is it against the law to sing? My voice isn't that good, but really!"

---

2. Lieutenant McCole declined my invitation to be interviewd, as did others in the state police. I therefore do not know who ordered McCole to try to get onto the site without first removing the farm equipment from the road. The sheriff strongly believes that someone who was not in the field gave McCole a direct order to walk around the barricades and march toward the site. He speculated that Superintendent Constantine might even have been in the communication trailer. Or he might have given the order from Albany. It is also possible that the order came from Major Kelley, who must have been in the communication trailer, because he wasn't in the field but later issued the official police press release. Though less likely, it is possible that McCole, who had shown signs of frustration with protesters at past confrontations, made the decision himself.

"People! Start hiking up to the site!" shouted Spike. "We'll have to make our stand there." Out of the corner of his eye, he saw a trooper walking toward him, about three feet away. He spun around and sprinted toward the van, where I was waiting. "Let's get the hell out of here!" he said for the second time that day.

As we sped off, I looked behind us and could see troopers arresting people who were walking slowly uphill toward the site. Spike slowed down when he reached the manure spreader, parked in a driveway near the fourth roadblock. Would he have time to get it positioned in the road, he wondered? What if he turned it on and filled the road with pig shit? A few seconds later, however, he sped by, doubting that he'd be able to get it in place before the police arrived. He also knew that if the police got covered in pig dung, it wouldn't play well on the six o'clock news.

"Damn! They're not even reading the injunction any more. I never liked wearing the damn mask anyway," he said more to himself than to me. Tearing it off, he defiantly turned to me as though I were responsible for the whole idea. "I'm not wearing it any more! We shouldn't have changed our strategy just because of the injunction."

I was plenty stressed, but the more ballistic Spike became, the calmer I felt. "Spike, we've got no choice but to wear the masks. Do you want to lose your house—even your horses—by getting identified?"

"The masks aren't stopping them," Spike objected. "We're going to have to use our bodies."

"Maybe so," I said. "But take it easy. Maybe when the cops see the huge crowds up at the site, they'll back off." We pulled into East Hill and Pinkerton where Stuart was the lone sentry.

Stuart Campbell had never felt more isolated. Ange and others were reporting that the troopers were marching around barriers and arresting people. A pickup truck pulled up alongside him. "You'd better do something," a woman shouted. "Everything's falling apart. The police aren't stopping for anything."

The truck roared by, heading toward the church. Stuart guessed there were at least five hundred people up there and at nearby roadblocks, some of whom he knew had been drinking. What would happen, he wondered, when they met the police? The crowd, he worried, could turn ugly. Someone would have to organize them, and quickly. Other people began streaming by Stuart on their way to the church, some demoralized, others furious. "This could be a disaster," Stuart said to himself.

He tried to raise Spike on the radio. "Sarge, this is Soupy." No response. Damn.

Stuart found himself reflecting on *Iphigenia*, a Greek play that he had studied with his students last semester in the Western civilization course. In Euripides' play, King Agamemnon had gathered together a huge Greek army to sack Troy in order to revenge Paris's abduction of Helen. But there was no wind and the ships couldn't sail. A priest from the temple of Artemis, goddess of nature, explained that the wind would not blow, because some of Agamemnon's soldiers had violated Artemis's grove and shot the sacred deer. The ships would remain stranded unless Agamemnon sacrificed his eldest daughter, Iphigenia, to the gods. And, after all, why shouldn't Agamemnon be willing to sacrifice his own daughter as the first victim of the war, when he was willing to risk so many of the best youth of Athens? Agamemnon vacillated. He tried to appease the army and sent for his daughter, believing that the wind would certainly return before she got there. Events, however, quickly stripped the king's power to command, and the army demanded his daughter's sacrifice.

Events at Caneadea were not going as planned. Stuart felt trapped like Agamemnon. Like the Greek king, ACNAG leaders had conjured up an army, which could be guided but no longer controlled. Unlike Agamemnon, however, things were falling apart too quickly for Stuart to have much time to agonize about making a decision. He'd have to organize the crowd now! But how?

"Horses!" he said to himself, instantly picking up the mike on his CB radio. "Maybe the horses can slow them down a bit until we get organized."

The horses were on the opposite side of the site. "Find Glenn," he said to someone monitoring a radio there. "Tell him to bring the horses down to East Hill and Pinkerton. Nothing else has stopped the police. They're marching around all the roadblocks, no longer bothering to read the injunction, and they've started arresting people." Hanging up the mike, Stuart realized he'd just broken his promise to Kay Eicher not to use horses in any confrontation. "Dammit," he said to himself, "the whole thing's dangerous. I can either let the police meet an angry mob, or send in the horses!"

Breathless, Sally Campbell spoke over the radio. "Stuart, I heard what you said. I was just down there. It's a rout. Nothing's going as planned. The horses are the only thing that can stop them. I ran as fast as I could, taking shortcuts across fields. I'm now at the intersection between you and the church. There are about sixty people here, but more are arriving from below all the time."

"Sally, I'm glad you're there. What's the mood?"

"Adrenalin is certainly flowing, but there's no panic. From what I can tell, the arrests have slowed the police down a bit. An officer has to walk each person a mile back down the hill to the police van."

"The horses may slow them down a bit more," Stuart said. "Prepare everyone to move down to this corner where we'll make our stand."

"Drew, are you there?"

"Yeah, I heard what you said to Sally."

"Get your people ready, too. Don't either of you move anyone yet. But get ready! When the time comes, we'll put everybody in the road. Right here there are open fields, and the police could just go around us. But about two hundred yards toward the site the road goes through a thick woods. That's where we should put everyone."

"Do you want me to bring my people down too?" asked Klaus Wuersig, a husky German immigrant who taught engineering at Alfred State College. He was a monitor at an intersection a half mile on the other side of the church.

"How many people are with you?" Stuart asked.

"About forty or fifty."

A plan had begun to form in Stuart's mind. He'd been worried that the police might just keep going straight up East Hill Road until they outflanked the protesters on the other side of the site. They could even bring their support vehicles around there.

"Klaus, when Sally and Drew move their people down to the north side of my intersection, you move yours around to the west side. That'll keep the cops from coming in the back door.

"All of you. Hold your positions for now, but get everyone ready to move. We'll see what happens with the horses."

A runner found Glenn Zweygardt and told him that Stuart wanted the horses. A large, muscular man, Glenn had grown up on a farm in Kansas where he'd helped plow fields with draft horses. He was now a sculptor whose gigantic metal and stone works referenced both the tectonic forces of nature and the gentler lakes, rivers, and hills of the Allegany County countryside. He was one of the few professors at Alfred University who mingled easily with county people at foundries and horse traders' conventions. Even better, Glenn had been in the dump fight from the beginning, first co-chairing the Alfred CCAC chapter, later heavily involved with ACNAG.

Glenn had not thought about strategies for using the horses. He certainly hadn't anticipated riding one. That morning he'd talked to the riders, none of whom seemed to know the plans. He assumed that Stuart had sent him to this outpost because he'd raised and trained horses. But no one had given him any instructions about using them.

The call from Stuart shocked him into realizing that he'd just been put in charge of the whole operation. The horses were saddled and waiting. "I guess I need a horse," he said.

"Here take this big guy. He's gentle." Someone showed him a huge Belgian. Although he'd ridden bareback for fun on draft horses when he was a child, he'd never seen one with a saddle before. This would be like driving a tank! Although he was a tall man, he couldn't quite reach the stirrup and someone had to help him up. The other eleven riders quickly mounted.

"Let's go!" one of them said. "We've got to get down there fast."

His own heart pounding, Glenn could feel the shared urgency. But he knew he'd have to calm himself down—and everyone else too. "Wait a minute. Take it easy. We've got a long way to go. Let's not run the horses and get them all goosey. We've got time, and we need to figure out what we're going to do."

Everyone pulled up around him. "The most important thing is to stay on your horse. Don't get pulled off. Don't ever let go of the reins, no matter what. Keep together. Stay bunched up and ride as a unit. Take it slow and easy."

No one challenged him as he took command of the group. The riders walked the horses south toward East Hill Road. Glenn had no idea what he'd be encountering and hoped that Stuart or someone else would give him a more detailed plan.

As soon as the riders turned onto East Hill Road, the police helicopter buzzed them. Glenn worried that the noise would spook the horses. The three saddle horses reared up and danced a bit, but the draft horses, used to working around machinery, appeared unaffected. "That guy's going to crash," yelled one of the riders. "Is he crazy?" cried another.

With the helicopter rumbling above them, they approached Stuart's intersection just after Spike and I had arrived. The riders halted. "Glenn, no one knows how to handle horses better than you," Stuart said. "Whatever you can do to slow down the police, do it. Stop them if you can."

"We'll see what we can do." Glenn now realized he was on his own.

"Who can give me a horse?" Spike asked. "I need to go down there." When no one volunteered, Spike told Glenn to pull him up behind him. Knowing that Spike was in charge of the site, Glenn gave him a hand up, even though he didn't like the idea of riding double into an encounter with the police.

They started down the hill. One of the saddle horses nervously pulled ahead as the helicopter dove toward them almost at the height of the electric wires. "Slow and easy," Glenn shouted above the loud drumming. "Let's keep together." As the horses walked slowly down the road, Spike told Glenn about police marching around roadblocks and arresting everyone in sight.

The riders encountered clusters of protesters rushing to get up to the church. When they saw the horses, many broke into applause. One woman

Mounting Civil Disobedience © Steve Myers 1990; all rights reserved.

imagined herself in a Western movie and yelled, "Hurray for the cavalry!" But it was a strange cavalry. The horses seemed to come from a mythic land—not only were they fabulous animals, but they were joined to legendary beings with yellow-orbed faces.

Soon, from the top of a hill, Glenn saw a long column of troopers three hundred yards in the distance. "Pull 'em up! Pull 'em up!" Glenn said, wanting to give the riders final instructions. The helicopter had momentarily departed and he could say a few words without shouting. He knew that charging the police lines would be too dangerous—and, of course, ACNAG couldn't then claim nonviolence. There could be only one strategy. The horses would have to slow down the police by milling around in front of them. He hoped the riders were experts.

"We've got to stay tight," Glenn said. "Stop in front of the troopers. Then circle the horses around in the road. Keep the reins up so the police can't grab them and pull you off. Don't let go of the reins for any reason. You've got to control the horses."

"Yeah," Spike agreed. "And keep the horses' rumps toward the police."

"That'll make it hard for the police to grab the reins," Glenn added.

"It'll also make it harder for the police to get around us," said Spike.

"Let's practice it now," said Glenn. The horses made a tight circle in the road, then turned uphill, moving from one side of the road to the other. "Great. Stay calm when we get down to the police and do the same thing." He turned and the horses walked toward the troopers, now two hundred yards away. The helicopter returned, frightening one of the saddle horses, which bolted ahead of the others. "Pull 'em up! Pull 'em up!" Glenn yelled. The rider got the horse under control about fifty yards from the police. The other horses walked up and they circled around in the road. A few protesters were singing "God Bless America," terribly offkey.

The troopers hesitated. Lieutenant McCole, looking dazed, walked into the midst of the horses, now walking slowly uphill. Glenn saw a trooper who had followed McCole into the mass of horse flesh jabbing a horse in the belly with his baton. Didn't he know that a horse could shatter his shinbone with one kick, Glenn thought. "I wouldn't do that," yelled a protester who was near the action. A couple of more troopers struck horses with their batons. "What the hell do you guys think you're doing," someone else shouted. "Those horses could kick the shit out of you!" Glenn was amazed at the Belgians' restraint.

Three troopers suddenly pulled Fran Root's twenty-eight-year-old son, Carl, from his horse. While Root struggled to hold onto the reins, an officer grabbed him around his neck. "Choke him out! Choke him out!" yelled a trooper a few feet away. The officer who had grabbed him around the waist screamed in pain. "Oh Jesus, my foot! Get the horse off my foot!" Wincing, he limped down the road where most of the police still waited. An officer shouted from the sideline to the men trying to separate Root from his horse, "Break his hand off the rein!" Two troopers then battered Root's hand with billy clubs until he finally let go.

Sitting precariously on the horse's rump, Spike realized his vulnerability, "Glenn, get me the hell out of here!" Still a bit irritated that Spike had thought it necessary to ride behind him into the action, Glenn took him up-hill and dropped him off.

A crowd of more than fifty people had gathered on the sidelines. Troopers were ruthlessly shoving people with cameras away from the action, but camcorders were everywhere. As the trooper pulverized Root's hand, the crowd became frenzied. "What the hell are you guys doing?" "You're New Yorkers too." "Police State!" A woman whose voice sounded like chalk rasping across a blackboard sang "We shall not be moved." And the wife of a farmer sobbed hysterically, crying out, "Stop this! Stop this!" She ran out toward Root until a trooper restrained her.

As Glenn returned, he saw a police officer strike one of the three saddle horses and grab the rider's reins as the startled horse turned around. Pulling the twenty-three-year-old rider down, an officer placed his foot on the young man's neck and ground his face into the mud. Like Carl Root, he was charged not only with resisting arrest, but also with a felony—second degree assault against a police officer.

The crowd grew increasingly angry when they saw police drag the two riders from their horses. "Police brutality!" someone shouted. A masked woman, daring police to pull her off her horse, turned it toward them and said, "What do you think you're doing! You should be ashamed of yourselves!" A bystander, wearing a sports jacket and tie, engaged McCole, telling him that he'd seen one of his officers strike a horse and beat the rider. "Have the trooper raise his hand, so we can identify him later."

"I'll do even better than that," snapped the lieutenant. "You can follow me back to headquarters and make an official complaint." Most of the troopers huddled a few yards downhill, as the nine remaining riders regrouped and circled their horses in the road. Glenn, usually an easygoing man, had a stubborn streak. And now he was angry. The troopers, he vowed to himself, would have to take down every last rider before they'd get by.

McCole radioed his superiors. "We can do it. We can get on the site. But I don't see the sense in it. It could turn into a very ugly situation." This time his bosses apparently relented, and shortly before 2 P.M. the police retreated, still more than a mile from the site.

Thirty-nine people had been arrested, including the six grandparents at the bridge and the two horsemen. All were charged with disorderly conduct and faced Judge Gorski's wrath for violating the injunction.

❖

No one within twenty-five miles of the Caneadea site could make a phone call that day—the phones had gone dead early in the morning. Newspaper reporters therefore had to leave the area to call in their stories. Only Kathryn Ross of the *Wellsville Daily Reporter* was on the scene at the beginning of the dramatic encounter between horses and police. Nor were any TV cameras there. That night newspapers and television stations relied heavily on a police statement that told of horses charging their lines, and recklessly endangering everyone. Even the *New York Times* printed the police version without checking other sources.

Frustrated that she couldn't get ACNAG's side of the story out to the press, Sally Campbell went home with her husband. Too exhausted to work on a press release, but too wired for sleep, she went to Peer Bode's media laboratory to review his students' videotapes. Luckily for the protesters, four of Bode's students had recorded the incident with the horses. That night Bode put them serially on one tape, then made dozens of copies. Assemblyman Hasper took one to the governor in Albany the next day. ACNAG also made tapes available to reporters, while simultaneously issuing a pithy statement: "1) No demonstrator, mounted or otherwise, perpetrated an act of violence. 2) The horsemen did not ride into police lines. 3) The police chose to enter the group of horsemen in order to continue escorting the technicians toward the site. 4) All injuries resulted from police pulling a rider from his horse. The rider was beaten and the horse stepped on a trooper's foot."

The use of horses, however, seriously divided protesters in Allegany County for the first time. Some agreed with the sheriff that they had changed a tense situation into a dangerous one. Some thought any use of horses was a violent act. Many more, however, agreed with the Campbells that the horses had prevented a bloodbath up on the plateau. The split was most serious among people who lived near the Caneadea site, where several people felt betrayed by ACNAG's leaders for using horses at all. The moral indignation of some created a backlash among many others who rejoiced that the horses not only had saved the day in Caneadea, but may also have given Allegany County its final victory over the siting commission. These people thought that ACNAG's leaders should be treated as heroes and found the strident criticism offensive.

Exhaustion and frayed nerves exacerbated tensions. Hundreds of people in the county had put their lives on hold as they organized rallies, lobbied legislators, made quilts that depicted the county's defiance, sold raffle tickets to raise money, and agonized about nonviolent strategies. Many struggled to pay huge monthly phone bills. Some businesses failed and others floundered as

Warning to State Troopers © Steve Myers 1990; all rights reserved.

people devoted themselves to the resistance. Though the struggle strengthened a few marriages, it strained others, a few ending in divorce. In the last year police had arrested one hundred twenty-nine people in the county, only a handful more than once.

For the first time in the struggle, citizens criticized the state police. Protesters put yellow and black road signs on county roads, with a circle and a diagonal line across a trooper beating a horse. People wrote scathing letters to the local papers. The most poignant came from Dennis Butts, whose brother had recently retired from the state police. He wrote about his small daughter's comment, while she was watching the videotapes with him. "I'm glad," she said, "the police are there . . . because they won't let them bring in the 'bad dirt.'" Butts "shuddered," wondering whether he would later "have to explain why the police helped bring in the 'bad dirt.'"

Alex Landis made the most conciliatory statement, "I hope that the whole New York state police department is not dubbed with the slogan of 'police brutality' because many were in complete control of themselves." Her statement, of course, implied that a few were not.

ACNAG's leaders, however, did not subscribe to the theory that "a few bad apples" were responsible for the events. The police, after all, were trained to follow orders. The recipe for disaster had occurred when the brass in the command headquarters ordered the troopers to depart from the ritualized scenario. Of course the police would do whatever was necessary to get the siting commission on the land, when they were ordered to do so.

Comments by the head of the siting commission again helped unify protesters in outrage and glee, as had his mushroom comments a few months earlier. This time, Orazio charged that the leaders' "exploitation" of the grandparents was "unconscionable." He didn't "hold those elderly people responsible," of course, because others had driven them "into a frenzy of fear" and "scared them to death." He claimed that if they "really had the opportunity to know the facts as we do, and discuss the issue openly," they wouldn't "feel compelled to do what they did."

The six elders delighted the county by writing an open letter to Orazio. "Exploited! The very thought makes those who know us personally roar with laughter." They noted that three of the people had Ph.D.s—one in chemistry. The others were equally knowledgeable. "No one here is exploiting us, Mister Orazio, but you are trying to exploit us by writing us off as elderly victims of some sinister deception by our friends and neighbors in Allegany County."

Some weren't "roaring with laughter," however. Joan Dickenson, the ace reporter for the *Olean Times Herald*, voiced the outrage of many. "Old people aren't children. I saw six adults who knew what they were doing. Whether they're right or wrong isn't the point." She ended her personal column gratefully acknowledging their action: "Those six people are Allegany County, and they are wonderful."

Epilogue

> In December 1988, when county residents started organizing to fight the potential dump, "I wouldn't have given them a snowball's chance," he said. "Now they've won. It's just a question of time."
>
> —Gary Horowitz, an administrator and former professor of American history at Alfred University, in *Olean Times Herald*, April 12, 1990

ALTHOUGH IT WASN'T CLEAR to most county citizens at the time, their triumph in keeping the technical team off the Caneadea site marked the final victory for Allegany County. It also rekindled a healthy debate across the United States about storing nuclear waste. At the time of this book's publication, no new dump site for low level nuclear waste has been built in the United States. Twelve years ago, twelve new sites were actively being planned. Ironically, Governor Cuomo, who earlier had belittled the activists' scientific arguments, now validated them publicly.

Worried that future confrontations would imperil police and protesters alike, Governor Cuomo told the siting commission to discontinue its attempts to conduct on-site testing. Just two days after the technical team's failure to get onto the Caneadea site, the governor asked the commission to "concentrate its efforts on other more productive activities," until he had time "to discuss this matter further with state legislators and local citizens."

The governor, however, walked a tightrope. He had to certify that New York was making progress in building a nuclear dump,[1] even though he now found many of the activists' arguments compelling. On at least two occasions

---

1. The Federal Low Level Radioactive Waste Act of 1985 allowed the three states that then accepted low level nuclear waste (South Carolina, Washington, and Nevada) to stop accepting shipments of waste from any state that was not in compliance with the law. A state (or compact of states) was out of compliance if it did not meet *(continued next page)*

he told protesters that he was grateful to them for educating federal and state officials (presumably including himself) about the serious issues involved in storing low level nuclear waste, and he urged them to stay involved.

The "process put together by the federal government," he admitted, was neither "rational" nor "fair." He doubted that members of Congress "understood the problem well when it passed that legislation." In any event, Congress had "acted unreasonably and unconstitutionally in imposing upon the states the legal responsibility for solving our nation's low-level nuclear waste problem."

What did Cuomo find so objectionable? First, class C waste was far too dangerous to be categorized as low level waste; rather, it should be reclassified as high level waste and stored in a federal repository.[2] Second, there was no need, he said, "to proliferate low-level radioactive waste sites all over the country. There should be one place for all of it." Third, he agreed with the activists that Congress had no right to force the states to "take title" to nuclear waste.

Although the governor himself did not explicitly fault the siting commission's scientific process in selecting a dump site, he told activists that he would ask the National Academy of Sciences to review the technical data. If it found the process seriously defective, he would recommend that the finalist sites be excluded from further consideration. Cuomo also suggested that the commission consider on-site storage of nuclear waste, a proposal that Ted Taylor had recommended in a technical report three months earlier. It would make more sense, he claimed, to send the "so-called low-level waste" to the same federal repository where they would be putting "the big stuff." On-site

---

(1. *continued from previous page*) certain deadlines. Technically, all states, including New York, had failed to meet certain deadlines, but the three states with nuclear dumps agreed not to penalize any state that was making progress toward building its own facility. Even though Cuomo argued against the wisdom of the federal law, he didn't want to risk triggering the mechanism that would prevent New York's nuclear facilities from continuing to send their waste to one of the established dumps. Instead, he hoped to modify the New York State siting law to make it better, while continuing to argue that the federal government should revisit the issue and take responsibility for nuclear waste. After all, he said, the federal government had done all the licensing of the nuclear plants.

2. The most dangerous "C" class of nuclear waste would account for 0.7 percent of the volume in the proposed dump, but it contained 93.9 percent of the radioactivity. This waste included some materials with very long "half-lives"; a small amount would remain radioactive for hundreds of thousands of years. Even the siting commission, while minimizing its danger, never publicly defended the inclusion of "C" class nuclear waste in a low level radioactive waste dump, but said the classification system was established in federal law and therefore beyond their purview.

storage could, therefore, be a viable temporary solution to solving the low level nuclear waste problem.

The activists were elated. The governor seemed to agree with all their arguments. In May the co-chairs of CCAC, Jim Lucey and Sue Beckhorn, attended a meeting in Cortland County with the governor. Both were cautiously optimistic. Beckhorn, usually very suspicious and critical of government officials, told the press that she "felt good that Cuomo's willing at least to listen. . . . The very fact that on-site storage—at the nuclear power plants—is part of his vocabulary now is encouraging. A few months ago, nobody was talking about it."

This optimism, however, gradually turned to disappointment. The reforms the governor sent to the legislature, though important, failed to address some of the citizens' most critical concerns. It did not make on-site storage a priority, and it did not challenge the siting selection process.

It did, however, require the siting commission to select a method for storing the waste, and to have that method certified through an environmental impact statement, before "continuing" the process of site selection. That would mean the citizens of Allegany and Cortland counties would get a "breather" of a year or two before they would again be facing the siting commission. (Assemblyman Hasper noted that the delay would also take the governor past the deadline for deciding whether he would run for president.) The other major change removed language about the state taking title to the nuclear waste.[3]

The governor said he would propose this legislation, however, only if the protest leaders and legislators of Allegany and Cortland Counties supported it. The governor's aides had consulted and bargained with activists in both counties and now the governor expected their support. Jim Lucey had helped forge the compromise in consultation with attorney David Seeger. Both pressured CCAC to actively support the governor's bill. At an extremely divisive CCAC meeting they noted that the legislation, while not perfect, could radically change the siting process in New York. The legislature had to remove the "take title" provisions from the 1989 state legislation or New York would end up owning all of the nuclear trash. They argued that a "certifiable" method for storing the waste would require the commission to face hard issues, such as the insane notion that "C" class nuclear waste could be included in a "low level" dump.

---

3. The other changes seemed like window dressing to most protesters in Allegany County. The siting commission would be expanded from five to seven members with the addition of a psychologist and an environmental activist. The citizens advisory committee would gain autonomy from the siting commission, and could theoretically become a forum for critical examination of the commission's work.

The depth of opposition at the meeting surprised the two men. ACNAG's leaders attended, arguing that the county had succeeded so far because people had not compromised. No officials in Allegany County should therefore endorse any bill about building a nuclear dump, especially a bill that left the commission's site selection in place. The fight had not been about how to build a "better dump," but about preventing the county from becoming a dumping ground for nuclear waste. Stuart Campbell concluded his statement, "If either CCAC or the county legislature endorse legislation that leaves us targeted, then we deserve to get the nuclear dump!"

Sue Beckhorn confronted Jim Lucey, her friend and co-chair of CCAC, in the most dramatic moment of the meeting. "This is what I think of the governor's proposal!" she said, throwing down a stack of papers on the floor. "He's stalling. Nothing's changed. All they're doing is postponing and delaying in an effort to get us to work with them. They're trying to manipulate us into helping them select a site." Becoming very emotional she added, "I haven't gone through these fifteen months of hell to have us throw it away now!" The vote of people present at the meeting overwhelmingly supported Beckhorn's position.

Allegany County's legislative leaders followed CCAC's lead and also refused to endorse the governor's legislation. Protest leaders and legislators in Cortland County, on the other hand, enthusiastically supported the bill, and despite Allegany County's lack of support, the governor proposed it to the legislature. When newspaper reporters asked the governor why Allegany County did not support the bill, he sighed and said, "Who is for a dump in Allegany County? The protesters know I'm not. I don't want a dump anywhere. I'm dealing with a federal law and the law is the law. . . . My position is to kill the federal statute and to get a more intelligent federal law."

"Why do you suppose," a reporter asked, "the protesters don't understand your position?"

The exasperated governor shrugged, "A lot of them don't know my position because they're always singing."

Assemblyman Hasper and state senator Jesse Present, who represented Allegany County, then had to decide whether or not to vote for the bill. Most county leaders urged them to vote against it, because it changed little and their votes might be interpreted as support for the continuing siting process.[4] Though they did not use their influence to keep the bill bottled up in committee, both

---

4. The advice was not, however, unanimous. For example, an editorial (June 29, 1990) in the *Wellsville Daily Reporter,* the only daily newspaper in Allegany County, urged both men to vote for the legislation, which would "extend the half-life of the siting process. . . . Time. That's what the governor's low-level radioactive waste amendment bill

legislators voted against it, ironically joining a handful of pro-nuclear legislators. Allegany County continued its defiance. There would not be the slightest compromise. The bill passed with large majorities in both the Assembly and State Senate and not only helped delay the siting process, but even kill it.

Allegany County strongly supported another legislative action, however. Hasper, Present, and the legislators who represented Cortland County spearheaded a movement to strip the siting commission of its funding. The governor proposed an increase in the siting commission's budget to $19.3 million for fiscal year 1990–1991 from $5.87 million in the previous year. The final appropriation bill outraged members of the siting commission when the legislature did not increase their budget at all, effectively reducing the governor's request by 66 percent.[5] Commissioner Richard Wood called the action "demoralizing. I don't see how we can go forward in a professionally responsible way, under these conditions."

When the governor signaled that the siting commission should stop its efforts to do technical tests on the sites, he removed all the steam from the legal boiler. In early June, Allegany County's district attorney dropped the "disorderly conduct" charges, because he couldn't "see where justice would be served by running an almost parallel procedure"; the protesters, after all, were facing charges for violating Judge Gorski's preliminary injunction.

The thirty-nine arrested people had already appeared once before Judge Gorski on May 19, 1990. As the first order of business, the judge, in agreement with the attorney general's office, dismissed charges against Sue Beckhorn and Fleurette Pelletier. Videotapes clearly showed the two women singing in a field off the road when the police arrested them. Jerry Fowler then made a series of motions and the judge postponed proceedings to consider them. Three motions were central. First, Fowler asked the judge to remove Assistant Attorney General Peter Sullivan from prosecuting the case, because videotapes clearly showed him talking to police during the action at Caneadea. This, at the very least,

---

will buy Allegany County." Although the editor could "understand why the Allegany County Concerned Citizens would reject it," he urged CCAC to have "faith in its own ability to improve and win the war, as it has the many battles and skirmishes." The editorial then enumerated several "fingerholds and toeholds in the slate-like wall of the siting commission" that the legislation would provide.

5. The reduction in funds for the siting commission was even more severe than these figures suggest, because the commission actually spent approximately $8.87 million in fiscal year 1989–1990, because they had supplemented the budget with $3 million in unspent funds from previous allocations.

made him a material witness and might even have made him a decision maker. Second, he asked the judge to proceed against each arrested person individually; the mere presence of someone at the action, even if the person wore a mask, did not show that the individual had violated the injunction or even knew about it. Third, Fowler asked the judge to vacate the injunction, because the siting commission had used it "to suppress opponents by seizing thirty-nine prisoners for contempt prosecution," rather than its intended purpose "to facilitate scientific testing."

A few days later the judge ruled on these motions. Although he refused to vacate the injunction or remove Sullivan from the case, he did suggest that the assistant attorney general "make a careful evaluation of his participation in the trial of this matter." Fowler believed that the judge's comments were a polite way to tell Sullivan that he had compromised himself by being present at Caneadea, though he reserved judgment about how serious the breach had been. Most significantly, the judge agreed with Fowler that each case had to be tried individually.

Having viewed the Caneadea videotapes, ACNAG's leaders were confident that most of the arrests were questionable. Aside from the six "grandparents," the others seemed arbitrary, since the protesters were not actually blocking the police. In order to build cases against those arrested, Sullivan offered immunity to the "grandparents," if they agreed to testify against others in ACNAG. All six said they'd go to jail before they'd cooperate. Always outspoken, Landis responded immediately. "There's no question in my mind. I can't let others take the rap for me. This thing is unjust!" Apparently the assistant attorney general had no stomach for sending elderly citizens to jail and could find no other way to build cases against those arrested. For whatever reason, the contempt cases just withered away in judicial limbo.

In August, Allegany County's district attorney presented the felony cases against the two horsemen to a secret grand jury, which, after viewing the videotapes and talking to witnesses, voted to dismiss the charges. They also voted not to indict any state troopers for excessive brutality. The D.A. later told the press that it was his "personal opinion that the grand jury wanted to pour water on the fire." The sheriff told reporters that the decision was "excellent" because it looked forward to conciliation rather than past to conflict.

New York State had led all other states in the siting process until it encountered the well-organized defiance of people in Allegany County.

Throughout 1990 Spike Jones visited every community with a proposed nuclear waste dump in the entire United States, helping to galvanize protests by telling Allegany County's story. When he arrived, people were often demoralized; when he left, people renewed their struggles and frequently engaged in civil disobedience. Repeatedly, he told people not to cooperate in any way with the siting process. "Don't even talk with them. You know enough not to talk to a car salesman if you don't want to buy a car, so why talk to a nuclear salesman if you don't want a dump?" Fighting nuclear dumps in the courts, he claimed, drained people's energies while making lawyers rich. People had to be willing to engage in civil disobedience if they were to win. He explained why nonviolence was essential: violence would split a community apart; nonviolence would clearly distinguish between the state as aggressor and the community as victim. But civil disobedience required committed individuals who were willing to risk jail and heavy fines to keep nuclear waste out of their communities.

"The New York style protests," as the nuclear industry dubbed them, spread across the United States. In some cases nonviolent resistance stopped the siting process outright, in others it significantly delayed it.[6]

New York not only provided the model for nonviolent resistance, it successfully challenged the federal Low-Level Radioactive Waste Policy Act of 1985 in the United States Supreme Court. Very early in the struggle, CCAC's lawyer, David Seeger, outlined the challenge, arguing that the federal law violated the Tenth Amendment to the U.S. Constitution because Congress lacked the power to force a state to "take title" of waste generated by private industry.

CCAC eventually convinced Governor Cuomo to file suit against the United States government. The governor considered the case to be reasonable, even though he was not overly optimistic, since the courts had in recent years

---

6. It would be too complicated here to detail what happened in each case after Spike's visit to an area. But here are two examples. Spike made a couple of trips to Connecticut. Activists in "Citizens Opposed to Waste" (COW) were receptive to his message and immediately initiated a civil disobedience action. The state immediately backed away from the proposed site to study the issues more fully. Nothing else of significance happened in Connecticut, and the siting process just withered away. People in Boyd County, Nebraska were also very receptive to Spike's message. As part of the Central Compact, this county would accept waste from Arkansas, Louisiana, Oklahoma, Kentucky, and Nebraska. The compact had hired a private firm to site the dump. When Spike arrived in June 1990, the company has already narrowed the selection to one site in Boyd County. The siting process in Nebraska was closer to completion than anywhere else, with the possible exceptions of Illinois and California. The people in Boyd County struggled mightily, engaging in several acts of civil disobedience. With difficulty they were able to slow down the process enough to keep the dump from being built. Eventually the federal law itself collapsed in the Supreme Court decision.

always sided with the federal government in Tenth Amendment cases. The U.S. District Court for the Northern District of New York dismissed the case and the U.S. Court of Appeals for the Second Circuit affirmed the lower court's decision, thereby validating Cuomo's pessimism.

The U.S. Supreme Court, surprisingly, heard the case on March 30, 1992. In a six to three decision, the justices declared the "take title provision" of the federal law "was unconstitutional."[7] Even though the Court upheld the rest of the act, including the "carrot" of monetary inducements, it struck down the "stick" that would force the states to comply. In the months after the decision, the siting process in most states collapsed.

When, in April 1995, the governor of South Carolina agreed to let Barnwell continue to accept low level nuclear waste indefinitely, pressure on the states to build their own dumps almost completely evaporated.

In 1996 the National Research Council (NRC) issued a scathing report (albeit in polite, scientific language) about the siting commission's site selection process in New York State. Governor Cuomo had asked the NRC to provide an independent panel of experts to evaluate the siting commission's work. The 1990 amendment to New York State's LLRW Disposal act required an independent scientific review. The NRC report involved several hundred pages of technical discussions, criticizing many details of the siting process.

John Hasper's technical committee, chaired by Gary Ostrower, had criticized the siting commission on several grounds. First, the county report criticized the siting commission's use of statistics in determining the "Candidate Area Identification" or CAI. In 1996 the NRC report concluded "that there is not a strong correlation between a score and the likely performance of an area as an LLRW disposal facility and, therefore, . . . the Siting Commission's use of cutoff scores in the CAI process was inappropriate." And during the "Potential Sites Identification" screening process, "the Siting Commission should have performed a detailed sensitivity analysis to test the effects of preference criteria weighting and scaling factors on scoring." The NRC's independent analysis showed that

---

7. The court did not agree with New York that the entire 1985 federal law was unconstitutional, however, just the "take title provision." The court argued this provision was unconstitutional, "because (a) an instruction to state governments to take title to waste, standing alone, would be beyond the authority of Congress, (b) a direct order to regulate, standing alone, would also be invalid, and therefore (c) Congress lacked the power to offer the states a choice between the two."

there was a "weak correlation between a site's score and its suitability for LLRW disposal."

Second, Hasper's technical committee had criticized the siting commission for using inadequate data in selecting the candidate areas. The NRC report concluded, "the results of screening were biased toward rural areas and—a problem of particular concern—regions that lacked data."

Third, Ostrower had repeatedly reproached the siting commission for ignoring data about oil wells, aquifers, and earthquake faults in Allegany County that the technical committee had sent. Nor had they revised any of their initial scores based on the new information. The NRC report concluded that one "cause of failure was the lack of a . . . quality assurance program." Such a program "would have provided a feedback mechanism to allow corrections to the siting program in a timely manner."

Fourth, the experts on Hasper's technical committee repeatedly criticized the siting commission for its unwillingness to share its primary data and to explain the basis for many of its decisions. The NRC apparently had the same problem, concluding that many decisions of the siting commission were "poorly documented and had a number of technical shortcomings." The NRC noted "that the Siting Commission did not provide adequate documentation of some of its important screening decisions." (Perhaps the problem was not that the siting commission wouldn't share their data; maybe they didn't have any to share.)

Though an interesting document vindicating many of the activists' technical criticisms of the siting process, the NRC report came far too late to affect New York State's siting process.

While the siting process in New York State slowly ground to a halt, Alex Landis's flag took on a symbolic life of its own. The sheriff had kept the flag in his office for more than a month after the action at Caneadea. He had hoped that Landis would change her mind and take it back after everything cooled down. When she did not relent, he sent it to Governor Cuomo on May 16.

One week later, Landis was sitting in her kitchen when the phone rang. "Hello. Is this Miz Landis?"

"Yes."

"This is Governor Cuomo. I'm sorry that you felt the government let you down. I'd like you to give us a chance to restore your faith in government."

"The siting commission came into Allegany County in an arbitrary way, with no consideration, no concern for the things the flag stands for."

"We're going to try to change that. I've been working with people from your county and Cortland County, and we're putting together some legislation that will change things." The governor then explained how the federal law had tied the state's hands and explained some of the things he was proposing. "I suggest you take back your flag and see what happens."

Landis agreed. The governor had the state police drive the flag back to the sheriff's department the next day, so that she would have it for Memorial Day.

Even before Landis got her flag back, she regretted her decision. She should have told the governor to send it to President Bush, since the federal government had not taken any steps to rethink its responsibility. So Landis herself sent the flag to the president. The federal government, refusing to accept it, returned it by mail. "That's typical," Landis told people in the county, "of the federal government's failure to accept responsibility for the nuclear waste. As long as the flag no longer stands for liberty, we'll have to continue opposing injustice."

Sources

THUCYDIDES, THE FATHER OF HISTORY, set out to write the true story of the Peloponnesian War. He would not include any fabulous tales. He said that "only after investigating with the greatest possible accuracy each detail" did he include anything. But his goal was to capture the motives of the men who planned and executed the war by using their words and speeches. He admitted that "it has been difficult to recall with strict accuracy the words actually spoken, both for me as regards that which I myself heard, and for those who from various other sources have brought me reports."

Even though these words and speeches of the principal actors in the Peloponnesian War were not recorded, Thucydides believed that the people's style of speaking and their words would allow the readers throughout history to understand what they thought and what motivated them to take the actions they did. I also believe that the words and style of speaking of the participants in Allegany County's fight against the nuclear dump give a much richer texture to the story. Like Thucydides, I have found that the people's words and conversations allow me to present a greater variety of views directly from the participants. The dialogue helps make the story readable, and, equally important, allows the participants to speak directly to the reader without the author's filter and therefore more accurately to present their motives.

Thucydides found that collecting all the information and making judgments about historical accuracy was, in his words, a "laborious task." *I certainly concur.* But I had an advantage over Thucydides insofar as I live in a mass-media age. Especially important to me have been newspaper accounts and videotapes. With the help of others, I collected, photocopied, and arranged chronologically well over 1,500 articles from nine newspapers—most from the *Wellsville Daily Reporter* and the *Olean Times Herald*. (Only after I had finished this task did I learn that Pam Lakin had collected and photocopied newspaper clippings for Herrick Library's special collection. Her collection is now housed there and a duplicate is in the county historian's office in Belmont.) I have about two dozen videotapes of meetings and actions. For example, there were six videotapes of the last civil disobedience action at Caneadea. In addition, I interviewed fifty people and made transcripts of their statements. (There are forty-one separate transcripts, however, because I interviewed several couples together.) These generous people who allowed me to interview them have an asterisk next to their name in the Index of Names. A few of them are not characters in the book, but they all provided important information about the events. The transcripts average around forty-five pages each. I sent each transcription back to the person whom I interviewed and asked them to make corrections or additions. All responded. I have placed duplicates of the videotapes and transcripts in both Herrick Library's special collection and in the office of the county historian in Belmont.

This book is nonfiction. I have invented no characters and I have changed no names. None of the participants requested anonymity, though I gave them that option. All the facts and dialogues are based on firm evidence. In most cases the quotations are taken directly from newspaper accounts or from videotapes. A few dialogues have been reconstructed from people's memories. But the words have been edited. At first I put in ellipses to indicate omissions. But that made the text awkward. It was also problematical, because newspaper reporters do not put people's quotations in chronological order and I felt free to rearrange them to make the story flow more naturally. Several professional historians with whom I have consulted agree with my decision. All history, after all, is edited. Some facts are reported; others are not. Some episodes are considered critical to understanding an event; some are ignored. I used the same spirit in editing the dialogues. I omitted nothing that the person would find critical in explaining his or her position. At least one-half of the episodes and dialogues have been read by the participants, who made a few minor revisions.

One of the hardest decisions was the style of referencing the work. One reader suggested that I footnote every sentence that was quoted and every motivation of the participants. Although that would be possible, since there is no quotation or motivation that is not based on evidence, it would have been a momentous task. It also would have doubled the length of the book. With the concurrence of editors at SUNY Press, I decided to cite the sources used in reconstructing each section of the book, often with a brief explanation about how the sources were used and for what purposes. Again, to keep the book manageable in length, I have cited only the newspaper's name and date, rather than the individual titles and bylines of the reporters. After a major action, for example, a single newspaper might have a dozen separate articles about the action, many by the same reporter.

There are twenty explanatory footnotes in the text, one-third of which are in the epilogue. Books and all other sources are included in this section.

ABBREVIATIONS OF NEWSPAPERS

BUF	*The Buffalo News*
D&C	*Rochester Democrat and Chronicle*
HET	*Hornell Evening Tribune*
OTH	*Olean Times Herald*
RTU	*Rochester Times Union*
SG	*Star Gazette* (Corning and Elmira)
SPEC	*The Sunday Spectator* (Hornell and Wellsville)
NYT	*New York Times*
WDR	*Wellsville Daily Reporter*

PROLOGUE

pp. 1–7

The dialogue came from the transcripts of Larry Scholes and Bill Timberlake. Corroboration by sheriff's comments to newspaper reporters at the time are found in OTH 12/12/89, OTH 12/15/89, OTH 12/26/89. The section about the meeting with the troopers from Albany was read over the phone to Larry Scholes, Bill Timberlake, and James Euken for any corrections. I made one small correction. The sheriff stated, "I can't swear every word is exact, but I have no corrections to make; I got goosebumps when you read it, because you sent me back there."

CHAPTER 1

pp. 9–12

The dialogue came from the transcript of Steve and Betsy Myers. The article is from NYT 12/21/88.

pp. 12–13

The information and dialogue came from the transcripts or Steve and Betsy Myers and Sandy Berry. The formation of local CCAC groups is reported in

HET 12/28/88, WDR 12/30/88, HET 12/30/88, WDR 1/4/89, OTH 1/5/89, OTH 1/6/89, OTH 1/7/89.

pp. 14–15

The information and dialogue came from the transcripts of Steve and Betsy Myers and Sandy Berry. Rich Kelley's transcript corroborates what happened at the initial meeting in West Almond. Corroboration also came from a journalist's report of the West Almond meeting in OTH 1/5/89.

pp. 15–16

The information and dialogue came from the transcripts of Steve and Betsy Myers and Gary Ostrower. Subsequent telephone conversations with both parties clarified some puzzling issues, especially Ostrower's motivation for telling Betsy she could attend the meeting, but that Steve could not.

pp. 16–20

The information and dialogue came from the transcripts of Gary Ostrower, John Hasper, Steve and Betsy Myers, and Irma Howard, and from reporters' accounts in OTH, 1/7/89, WDR 1/8/89, OTH 1/9/89. I also have a copy of the letter from Edward Coll, president of Alfred University. Provost Rick Ott additionally clarified some of the university's motivations in a conversation I had with him, and he provided me with a memo he sent President Coll after the meeting. Ott wrote lengthy newspaper articles on the university's position in HET 10/4/89 and OTH 12/3/89.

pp. 20–21

The dialogue came from the transcripts of Gary Ostrower and Steve and Betsy Myers.

pp. 21–25

The information and dialogue came from the transcripts of Steve and Betsy Myers, Sandy Berry, Jim Lucey, Rich Kelley, and Glenn Zweigardt.

pp. 25–27

Information and dialogue came from the transcripts of Steve and Betsy Myers, Sandy Berry, and John Hasper and from reporters' accounts of the Almond Fire Hall Meeting in OTH 1/14/89, SPEC 1/15/89. The text of the letter to trustees that was signed by 112 faculty members is reprinted in WDR

1/26/89. Information on West Valley came from Ray Vaughan's transcript and
the following articles: NYT 10/29/89, D&C 11/5/89, D&C 1/30/90, D&C
1/31/90, WDR 2/1/90. The OTH ran a series of in-depth articles by Asso-
ciated Press writer Alan Flippin on West Valley: 4/22/90, 4/23/90, 4/24/90.
The most widely read source for knowledge about nuclear waste storage was
the newspaper reprint version of what became a paperback book: Donald L.
Barlett and James B. Steele, *Forevermore: Nuclear Waste in America* (New York:
W.W. Norton, 1986). A second book that was widely read in the county is
Ralph Nader and John Abbotts, *The Menace of Atomic Energy* (New York: W.W.
Norton, 1977). See OTH 1/21/89 for Alfred University's revised position on
opposing the proposed dump. Houghton College's anti-dump position is re-
ported in OTH 1/30/89. Later Joan Dickenson wrote two articles on
Houghton College's position, including the range of opinion among faculty
and administrators: OTH 10/31/89, OTH 11/1/89.

pp. 28–30

Information and quotations were from the transcripts of Steve and Betsy
Myers, Gary Ostrower, Gene Hennard, Irma Howard, from my memories,
from the report of the County Technical Committee, and from reporters' accounts in OTH 3/30/89, OTH 4/3/89, and OTH 4/4/89. Joan Dickenson
then wrote a series of seventeen articles on the various parts of the report that
appeared in the OTH on the following dates: 4/6/89, 4/7/89, 4/10/89,
4/11/89, 4/12/89, 4/13/89, 4/14/89, 4/15/89, 4/16/89, 4/18/89,
4/19/89, 4/20/89, 4/21/89, 4/24/89, 4/25/89, 4/28/89, 4/29/89. For the
siting commission's response to the report see OTH 2/15/90, OTH 2/16/90,
and OTH 2/24/90. The section on the cluster flies was researched and written by Irma Howard. The section that critiqued the statistical methodology
used by the siting commission was written by Marion Weaver and Scott
Weaver. The National Academy of Science would later validate the Weavers'
conclusions and chide the siting commission for poor statistical analysis. See
Epilogue. See WDR 12/20/89 as a good example of how the technical committee's report was used to solidify citizens' opposition to the dump.

pp. 30–35

Most of the quotations are directly from the videotape of the meeting that was
made by Peer Bode. The videotape also provided most of the visual information
included here. Almost all the transcripts I have collected include sections about
this meeting. Motivations, thoughts, and other information came particularly
from the transcripts of Steve and Betsy Myers, Gary Ostrower, John Hasper,
Sue Beckhorn, Stuart Campbell, Jessie Shefrin, and Pam Lakin. Newspaper

reports from WDR 1/20/89, WDR 1/24/89, OTH 1/25/89, WDR 1/26/89, OTH 1/26/89, WDR 1/27/89, OTH 1/27/89, D&C 1/27/89, WDR 1/30/89 were consulted and provided some information and quotations. They especially gave important perspectives about what was going on outside the meeting hall. Videotapes of Jean Kessner's reports on Maillie and Wood that are reported here are in the archives. Unfortunately they are not dated, but internal evidence shows that they were in late October and early November 1989.

CHAPTER 2

pp. 37–39

Information and quotations came from the transcripts of Gary Lloyd, Walt Franklin, and Steve and Betsy Myers.

pp. 40–44

The discussion of civil disobedience did take place on April 10, 1989. I have, however, included some important conversations from a couple of subsequent CCAC meetings that occurred over the next couple of months—most notably David Seeger's concern over the possibility that the state would approve a temporary storage dump that would become the state's de facto nuclear dump. My motivation for conflating these meetings is to keep the book from becoming overly tedious by detailing each separate meeting. Also, participants had a good memory of the issues and what was said, but not usually of the particular meeting. This is the only instance in the book where I have taken the liberty of conflating meetings or episodes. Information and dialogue was created from the following transcripts: Jim Lucey, Steve and Betsy Myers, Gary Lloyd, Stuart Campbell, Sue Beckhorn, Fleurette Pelletier, and Glenna Fredrickson. In addition, notes from Sue Beckhorn's journal helped me reconstruct some of the issues in this section. Interim storage remained a serious issue for months—see OTH 1/9/90 and OTH 1/10/90 for a particularly clear description of the issue.

pp. 44–48

The information and dialogue came from the transcripts of Gary Lloyd, Jim Lucey, Walt Franklin, Stuart Campbell, Sally Campbell, Jessie Shefrin, Graham Marks, Hope Zaccagni, and Jerry Fowler, and from my memories. I also used Walt Franklin's diary. I got some additional information in phone conversations with Dave Davis and Freddy Fredrickson.

pp. 48–50

The information and dialogue came from the transcripts of Jim Lucey, Gary Lloyd, Stuart Campbell, Sally Campbell, and Jerry Fowler, and from my memories.

pp. 50–51

The quotations and thoughts came from Jim Lucey's transcript. Some corroboration came from Hope Zaccagni's transcript.

pp. 51–56

The quotations and motivations came from the transcripts of Rich Kelley, Graham Marks, Jessie Shefrin, Spike Jones, Carol Burdick, Pam Lakin, and Hope Zaccagni.

pp. 56–60

Quotations and dialogue came from the transcripts of Jim Lucey, Mary Gardner, Glenna Fredrickson, Stuart Campbell, Sally Campbell, Larry Scholes, Eliott Case, Rich Kelley, and Gary Ostrower, from my memories, and from reporters' accounts in OTH 5/30/89, OTH 6/1/89, WDR 6/1/89, HET 6/1/89, and WDR 6/5/89. The draft of Hasper's statement had initially been written by Gary Ostrower and faxed to him in Albany immediately following the meeting with the county legislators at the courthouse.

CHAPTER 3

pp. 61–64

Dialogue, motivation, and feelings came from the transcripts of Mary Gardner and Glenna Fredrickson. The newspaper article and picture were in WDR 8/2/89.

pp. 65–69

Most of the memories and dialogue were reconstructed from Spike Jones's transcript. Some additional information, memories, and quotations came from Gary Lloyd's transcript. Spike remembered that a "captain" of the State Police had called him. He could not remember if it was Browning. Browning was, however, the "captain" who subsequently had several dealings with ACNAG, and it is reasonable to assume that he had called Spike that day.

Spike often talked with Gary about his army experiences. I only slightly edited the story as Spike told it to me.

pp. 69–73

The dialogue and all quotations came from reporters' accounts in BUF 8/3/89, WDR 8/3/89, OTH 8/3/89, HET 8/3/89. Motivations and the thoughts of the participants came from the transcripts of Steve and Betsy Myers, Sue Beckhorn, Fleurette Pelletier, and Mario Cuomo. Governor Cuomo suggested that I read the account of his mediation in a housing dispute between Jewish and African American citizens in order to understand his thoughts about compromise. I've therefore used Mario Matthew Cuomo, *Forest Hills Diary: The Crisis of Low-Income Housing* (New York: Random House, 1974), p. 200 and *passim*.

pp. 73–74

The information and quotations came from the transcript of Stuart Campbell and the following newspaper accounts: BUF 8/3/89, WDR 8/3/89, OTH 8/3/89, HET 8/3/89.

pp. 74–75

The information and dialogue came from the transcripts of Spike Jones and Sue Beckhorn. See OTH 8/20/89 for Sue Beckhorn's open letter to Governor Cuomo.

pp. 75–79

The quotations and information came from reporters' accounts in OTH 9/10/89, D&C 9/10/89, BUF 9/10/89, SPEC 9/10/89, NYT 9/10/89, WDR 9/11/89, OTH 9/11/89, WDR, 9/11/89, HET 9/11/89, OTH 9/12/89, WDR 9/12/89, HET 9/12/89. Corroborative information came from the transcripts of John Hasper, Gary Ostrower, Steve and Betsy Myers, Sally Campbell, and Fred Sinclair.

pp. 79–87

Newspapers reported the meeting with Cuomo on Nov. 17: HET 11/17/89, OTH 11/18/89, OTH 11/19/89, SPEC 11/19/89, OTH 11/20/89, *Syracuse Herald American* 11/19/89. Additional information about this meeting came from the transcripts of Jim Lucey, Ted Taylor, Steve and Betsy Myers, and John Hasper. Ted Taylor's notes in his daily journal were also helpful in reporting the dialogue.

Newspapers did not report on the February 17, 1989, meeting in Fillmore. Again Ted Taylor's notes in his daily journal were very important in reconstructing the dialogue. The transcripts of Ted Taylor and Steve Myers were also helpful.

Quotations, dialogue, and information in the section where Ted Taylor explained his transformation from pronuclear advocate to antinuclear critic came from Ted Taylor's diary, from John PcPhee, *The Curve of Binding Energy* (New York: Farrar, Straus & Giroux, 1974), pp. 8, 110, 120, and *passim*, and from transcripts of Steve Myers and Ted Taylor. The quotations are drawn most heavily, however, from a paper Dr. Taylor presented to the Nuclear Dialogue Project Meeting, Princeton, N.J., on October 21, 1987: Theodore B. Taylor, "From Bomb Designer to Disarmament Activist." During the nuclear dump fight, Dr. Taylor used this paper as the basis for many presentations when he explained the reasons for his change of heart—the first during the meeting in Fillmore where Gary Ostrower gave the county's reports to members of the siting commission (see chapter 1, above). Articles on Ted Taylor and his analysis of the nuclear dump in local newspapers include OTH 5/23/89, SPEC 6/4/89, OTH 9/30/89, OTH 10/5/89, OTH 10/6/89, OTH 10/7/89, OTH 11/19/89, OTH 11/20/89, OTH 3/5/90, WDR 3/9/90.

CHAPTER 4

pp. 89–92

The information and dialogue came from the transcripts of Spike Jones, Gary Lloyd, and Stuart Campbell.

pp. 92–95

The information and quotations came from ACNAG's "Training Manual for Civil Disobedience," from a videotape of an ACNAG training session in Belfast, from the transcripts of Spike Jones and Sally Campbell, and from my memories. I set up and organized the c.d. training sessions. With Marion Kurath-Fitzsimmons and Sally Campbell, I also coordinated the writing of a manual that explained ACNAG's rules and suggested ways to keep the protests nonviolent. At an ACNAG steering committee meeting, people volunteered to become trainers. I called the group together and we discussed how we would run the training sessions that are described in this section. Joan Dickenson wrote a series of articles about ACNAG training sessions in OTH: 2/17/90, 2/18/90, 2/19/90, 2/20/90. Other newspaper reports include HET 9/25/89, OTH 10/2/89, WDR 10/4/89, WDR 12/21/89, WDR 1/28/90, WDR 2/16/90.

pp. 95–97

For initial reactions of the people in Allegany County to the naming of the three potential sites see OTH 9/10/89, D&C 9/10/89, BUF 9/10/89, SPEC 9/10/89. For CCAC's early legal work with landowners see OTH 9/15/89, WDR 9/20/89, OTH 9/20/89. Some background information also came from Jim Lucey's transcript.

pp. 97–102

Information and some quotations came from the transcripts of Mick Castle and Gary Lloyd. Reporters' accounts in WDR 9/22/89, HET 9/29/89, WDR 9/29/89, OTH 10/11/89, HET 10/12/89, WDR 11/17/89, HET 11/20/89, OTH 12/13/89, BUF 12/17/89, WDR 12/20/89, BUF 1/14/90 also supplied quotations and background information.

pp. 102–108

The quotations and information came from reporters' accounts in WDR 10/25/89, HET 10/26/89, RTU 10/27/89, WDR 10/27/89, OTH 10/27/89, and from the transcripts of Sue Beckhorn, Howard Appell, Glenna Fredrickson, Mary Gardner, Jim Lucey, and Rich Kelley. The numbers at the Night of Rage have been disputed. Estimates were difficult, because the event took place in a wilderness area where the only lights were from a large bonfire, lighted pumpkins, and a few generated lights on a makeshift stage. There were no police estimates of the crowd; reporters estimated between 400 to 1,000. I am using 750 as a rough estimate.

The books that anti-dump leaders were reading by Ernest J. Sternglass, the speaker at the "Night of Rage," included *Low Level Radiation* (New York: Ballantine, 1972) and *Secret Fallout: Low Level Radiation from Hiroshima to Three Mile Island* (New York: McGraw Hill, 1981). Another popular book that also presented Sternglass's position is Leslie J. Freeman, *Nuclear Witnesses: Insiders Speak out*. Other books that anti-dump leaders were reading on the effects of low level radiation on health were John W. Gofman, *Radiation and Human Health* (San Francisco: Sierra Club Books, 1981) and Rosalie Bertell, *No Immediate Danger: Prognosis for a Radioactive Earth* (Summertown, Tenn.: Book Pub. Co., 1985). In addition, John Gofman gave several cases of his *Irrevy: An Irreverent Illustrated View of Nuclear Power* (San Francisco: Committee for Nuclear Responsibility, 1979) to be distributed to anti-dump leaders and libraries in Allegany County without charge. This list is compiled from names of books that the anti-dump leaders mentioned to me as important. Many of them were kept at the Vigil.

pp. 108–110

The dialogue and information came from Kathryn Ross's account in WDR 11/16/89, and from the transcripts of Sally Campbell, Spike Jones, Mick Castle, Peg Jeffords, and Larry Scholes.

CHAPTER 5

pp. 111–113

The information and quotations came from reporters' accounts: OTH 12/5/89, HET 12/5/89, WDR 12/6/89, HET 12/11/89, OTH 12/12/89. Stuart Campbell's transcript provided supplementary information.

pp. 113–115

The information and quotations came from the transcripts of Craig Braack, Spike Jones, Sally Campbell, from my memories, and from reporters' accounts in OTH 12/13/89, WDR 12/13/89, HET 12/13/89.

pp. 115–118

The information and dialogue came from the transcripts of Pam Lakin and Stuart Campbell. This section was also checked with Carol Burdick for her recollections.

pp. 118–119

The information and dialogue came from the transcripts of Stuart Campbell, Pam Lakin, Bill and Carla Coch. This section was also checked with Carol Burdick for her recollections.

pp. 119–120

The information and dialogue came from the transcripts of Larry Scholes and Bill Timberlake.

pp. 120–125

The information and dialogue came from the transcripts of Craig Braack, Larry Scholes, and Bill Timberlake. Larry Scholes's interviews in the press that are quoted here are from OTH 12/10/89 and OTH 12/12/89. While no one remembered the exact words that Dorothy Chaffee used in "telling off" Bruce

Goodale, all three remember that she vented her anger. Ms. Chaffee, however, gave an interview to OTH on 12/8/89, only a few days before this episode where she told the reporter what she wanted to tell the siting commission; I used these newspaper quotations in reconstructing the dialogue that Ms. Chaffee had with Goodale.

CHAPTER 6

pp. 127–131

The information and some dialogue came from the transcripts of Stuart Campbell, Craig Braack, and Bill and Carla Coch. Quotations of the dialogue between Coch and Goodale came from OTH 12/14/89. On the increased danger of low level radiation in the BEIR V Report, see WDR 12/19/89 and Buf 1/19/89. William Coch took out a full-page advertisement in the WDR 1/8/90 to publicize the health risks of low dosages of radioactive waste that came from the BEIR V Report; he signed the advertisement and used his title, "Medical Advisor to the Allegany County Health Department."

pp. 131–132

The information and dialogue came from the transcripts of Stuart Campbell, Pam Lakin, Bill and Carla Coch, and Gary Lloyd.

pp. 132–134

The information and quotations came from the transcripts of Stuart Campbell, Larry Scholes, and Walt Franklin.

pp. 134–139

Most of the information and dialogue came from the following newspaper articles: RTU 12/14/89, OTH 12/14/89, HET 12/14/89, D&C 1/14/89, WDR 12/14/89, WDR 12/15/89. Supplementary information came from the transcripts of Stuart Campbell, Sally Campbell, Spike Jones, Gary Lloyd, Larry Scholes, from my memories, and from a phone discussion with Freddy Fredrickson.

pp. 139–143

Most of the quotations came from the following newspaper articles: RTU 12/14/89, OTH 12/14/89, HET 12/14/89, D&C 1/14/89, WDR 12/14/89, WDR 12/15/89. Additional information and dialogue came from the transcripts of Larry Scholes, Stuart Campbell, Gary Lloyd, William Giovanniello, Sally Campbell, Spike Jones, and Craig Braack, and from my memories.

p. 143

The information and quotations came from the transcripts of Larry Scholes and Bill Timberlake.

p. 144

The information and quotations came from a phone conversation with Wadi Sawabini that I immediately transcribed from detailed notes.

CHAPTER 7

pp. 145–146

The information and quotations came from the transcripts of Larry Scholes and Bill Timberlake, and from reporters' accounts in BUF 12/23/89 and OTH 12/26/89.

pp. 146–147

The information and quotations came from the transcripts of Larry Scholes and Bill Timberlake.

pp. 147–149

The information and quotations came from the transcripts of Larry Scholes, Bill Timberlake, Bill and Carla Coch, Stuart Campbell, Spike Jones, and Jerry Fowler. ACNAG had telephone interviews with reporters after this meeting. See newspaper accounts in WDR 1/9/90 and OTH 1/9/90, which helped supplement the participants' memories.

pp. 149–154

The information and dialogue came from the transcripts of Sally Campbell and Stuart Campbell and from reporters' accounts in OTH 1/12/90, OTH 1/14/90, and the Magazine of BUF 1/14/90.

pp. 154–159

The information and dialogue came from the transcripts of Hope Zaccagni, Spike Jones, Larry Scholes, and Sally Campbell, and from reporters' accounts in OTH 1/16/90, WDR 1/16/90, OTH 1/17/90, WDR 1/17/90, BUF 1/17/90, HET 1/17/90, and from videotapes.

p. 159

The information and dialogue came from the transcripts of Mary Gardner and Jim Lucey. See also HET 1/22/90.

pp. 159–162

The information and quotations in the dialogue mostly came from videotapes and from reporters' accounts in OTH 1/18/90, HET 1/18/90, WDR 1/18/90, OTH 1/19/90, WDR 1/19/90. The transcripts of Larry Scholes, Gary Lloyd, Mary Gardner, and Jim Lucey helped me ascribe motives and fill in a few gaps in the newspaper accounts.

pp. 162–166

The information and quotations about the siting commission trying to get an office came from reporters' accounts in HET 1/1/90 and OTH 1/5/90. News reports that include quotations from the two men who finally refused to rent their building to the siting commission are in WDR 2/1/90 and OTH 2/1/90. John Barnett and Gary Fuller were owners of B & F Housing, Inc. Orazio's mushroom comment and peoples' responses are in OTH 1/19/90, OTH 1/20/90, OTH 1/22/90, OTH 1/26/90, OTH 1/27/90, OTH 1/30/90, WDR 2/1/90, WDR 2/2/90, OTH 2/4/90, WDR 2/7/90, WDR 2/14/90, WDR 2/20/90. The transcripts of Jim Lucey, John Hasper, and Stuart Campbell, and my memories also supplemented the newspaper reports.

The following four articles written by social scientists examine the community organizing that took place in Allegany County: David Kowalewski and Karen L. Porter, "Environmental Concern among Local Citizens: A Test of Competing Perspectives," *Journal of Political and Military Sociology*, vol. 21, Summer, 1993: 37–62; Steven A. Peterson, David Kowalewski, and Karen L. Porter, "Dumpbusting: Symbolic Politics or NIMBY?" *Polity*, vol. 25, no. 4, 1993: 617–631; David Kowalewski and Karen L. Porter, "Ecoprotest: Alienation, Deprivation or Resources?" *Social Science Quarterly*, vol. 73, no. 3, 1992: 523–524; and David Kowalewski and Karen Porter, "Mushrooms in Our Backyard: The Structure of Citizen Response to Environmental Threats," *Sociological Viewpoints*, vol. 7, 1991: 1–23.

pp. 166–170

The information and quotations came from the transcripts of Gary Lloyd, Sally Campbell, Stuart Campbell, and Spike Jones, from my memories, and from reporters' accounts in OTH 3/1/90, WDR 3/1/90, BUF 3/1/90, OTH 3/2/90.

pp. 170–171

The information and quotations came from reporters' accounts in WDR 3/5/90, OTH 3/5/90, WDR 3/6/90, WDR 3/7/90, OTH 3/7/90, WDR 3/8/90, OTH 3/8/90, WDR 3/8/90, WDR 3/9/90, SPEC 3/11/90, WDR 3/12/90, and from the transcripts of Sue Beckhorn, Stuart Campbell, and Walt Franklin.

CHAPTER 8

pp. 173–176

The most significant articles for the quotations and dialogue came from reporters' accounts in OTH 2/21/90, WDR 2/21/90, D&C 2/21/90, and WDR 2/22/90. Other accounts on the background of the injunctions and the legal maneuvering include OTH 2/1/90, OTH 2/3/90, OTH 2/5/90, WDR 2/5/90, WDR 2/6/90, OTH 2/6/90, OTH 2/9/90, OTH 2/23/90, WDR 2/23/90, WDR 2/27/90. The police officers' affidavits were written and signed by Lieutenant Walter DeLap of Batavia, Lieutenant Charles McCole of Olean, and Investigator George Brown of Batavia.

pp. 176–179

The quotations and dialogue came from the transcripts of Sally Campbell, Gary Lloyd, Larry Scholes, and Bill Timberlake, and from my memories. Some corroboration for what Scholes said at this meeting comes from WDR 4/2/90 and WDR 4/4/90.

pp. 179–187

The quotations, dialogue, and other information came from the transcripts of Spike Jones, Sally Campbell, Stuart Campbell, Gary Lloyd, Roland Warren, Mary Gardner, Jessie Shefrin, Peer Bode, Jim Lucey, and Hope Zaccagni, and from my memories.

pp. 187–189

The information, quotations, and dialogue came from Hope Zaccagni's transcript, from my memories, and from the visual information in the yellow-orbed mask itself.

pp. 189–190

The information and quotations in this section came from my memories, and indirectly from the transcripts of Hope Zaccagni, Stuart Campbell, and Sally Campbell. There is no question that everyone believed that Judge Gorski had made the comment about wanting fifty arrestees. In writing this book, I have traced the comment back to its source in Allegany County, who said that it was definitely true, but would not reveal the source in or near the attorney general's office. This was the only statement in all my interviewing that a source wanted to be kept confidential. So I am reporting the "fact" here as leaders in ACNAG believed it at the time, but cannot validate its authenticity.

pp. 190–192

The information in this section came from the transcripts of Stuart Campbell, Sally Campbell, and Spike Jones. The memories of what happened this night do not coincide as well as they did for all the other events in this book. Emotions were running very high. I reconstructed the dialogue and conversation based on the *agreement* in these three accounts. I am convinced that I have captured the sense of foreboding and apprehension that existed among all three central figures in ACNAG. I also checked a few facts with Helen Hutchinson over the phone.

CHAPTER 9

pp. 193–196

The quotations and information came from the transcripts of Stuart Campbell and Sally Campbell and from reporters' accounts in OTH 4/5/90, WDR 4/5/90, OTH 4/6/90. Also see WDR 4/2/90 and WDR 4/4/90 for articles about the sheriff's and the protesters' expectations about what would happen on April 5. These articles also discuss how the injunctions might be changing the ground rules.

pp. 196–198

The quotations in this section came from Roland Warren's transcript. Background information also came from reporters' accounts in WDR 4/6/90, OTH 4/6/90, HET 4/6/90. See SPEC 3/4/90 for the first article that mentions elders as a group who opposed the nuclear dump.

John Leax, a professor of English at Houghton College and one of the monitors at the bridge, wrote about his religious reasons for getting involved in the fight against the nuclear dump in *Standing Ground: A Personal Story of Faith and Environmentalism* (Grand Rapids: Zondervan Publishing House, 1991).

pp. 198–199

The quotations and information in this section came from the transcripts of Sally Campbell, Stuart Campbell, and Spike Jones, and from my memories.

pp. 199–205

The quotations in the dialogue between the sheriff and me came from Larry Scholes's transcript and from my memories. The other quotations in this section came from videotapes of the action and from reporters' accounts in WDR 4/6/90, OTH 4/6/90, SG 4/6/90, HET 4/6/90.

pp. 205–210

The quotations and information about Spike's and my dialogue came from Spike Jones's transcript and from my memories. The quotations of the sheriff and people came from videotapes. Larry Scholes's transcript is the source for his thoughts about the police decision to leave police vehicles behind and march the three miles to the site.

pp. 210–212

The quotations and information came from the transcripts of Fleurette Pelletier, Sue Beckhorn, and Spike Jones, from my memories, and from videotapes of the action. Reporters' accounts in WDR 4/6/90, OTH 4/6/90, SG 4/6/90, HET 4/6/90, BUF 4/6/90, OTH 4/8/90, WDR 4/9/90 helped corroborate this information.

pp. 212–214

The quotations and information came from the transcripts of Stuart Campbell and Sally Campbell.

pp. 214–219

The quotations and information about sending the horses into the action came from the transcripts of Glenn Zweygardt, Stuart Campbell, and Spike Jones. The quotations of Lieutenant McCole and others at the encounter between the police and the horses was captured on several videotapes, including one that was subpoenaed from the state police. Peer Bode was coordinating most of the videotaping that took place and turned over all the videotapes to ACNAG. Further corroboration comes from reporters' accounts in WDR 4/6/90, OTH 4/6/90, SG 4/6/90, BUF 4/6/90, HET 4/6/90.

pp. 220–222

Most of the quotations and information came from reporters' accounts in NYT 4/6/90, WDR 4/6/90, OTH 4/6/90, OTH 4/8/90, WDR 4/9/90, OTH 4/12/90, WDR 4/12/90, WDR 4/20/90, WDR 5/29/90. Additional information came from the transcripts of Sally Campbell, Peer Bode, and Spike Jones.

No one I talked to had any explanation for the dead phones. The sheriff thought it had something to do with the heavy spring snow during the night. While most protesters found the situation suspicious, the sheriff denies that the police had anything to do with the outage.

The lack of phone service only partially explained the reporters' absence. Most of the TV reporters had left so they could get their tapes of the confrontation at the bridge back to their studios and have time to edit them for the evening news. The nearest TV stations are in Rochester and Buffalo, about one and one-half hours away from the Caneadea site. Wadi Sawabini, from the CBS affiliate in Buffalo, was still in the area, but had taken his crew up to the site itself to see how the protesters were organizing there for what he assumed would be the final confrontation. When he heard about the confrontation with the horses, he rushed his team down the hill in time to interview Scholes and McCole. He also made arrangements to use a citizen's videotape of the action. Only Channel 4 in Buffalo, therefore, played footage of the encounter between horses and police on their nightly news.

EPILOGUE

pp. 223–227

Most of the quotations and information came from reporters' accounts in OTH 4/7/90, BUF 4/7/90, D&C 4/7/90, SPEC, 4/8/90, OTH 4/8/90, OTH 4/11/90, WDR 4/23/90, WDR 5/1/90, BUF 5/1/90, OTH 5/17/90, OTH

5/24/90, D&C 5/24/90, WDR 5/24/90, BUF 5/24/90, WDR 5/31/90, OTH 6/13/90, WDR 6/14/90, BUF 6/14/90, OTH 6/28/90, WDR 6/28/90, WDR 6/29/90, SPEC 7/1/90, BUF 7/1/90, WDR 7/2/90, OTH 7/2/90, OTH 7/10/90, WDR 7/11/90, D&C 8/11/90, OTH 8/11/90, SPEC 8/12/90. Additional information came from the transcripts of Stuart Campbell, Spike Jones, Sue Beckhorn, and Jim Lucey, and from my memories.

pp. 227–228

The information primarily came from reporters' accounts in OTH 4/16/90, OTH 5/15/90, WDR 5/18/90, BUF 5/19/90, OTH 5/19/90, SPEC 5/20/90, OTH 5/30/90, WDR 5/30/90, OTH 6/6/90, WDR 6/6/90, WDR 6/20/90, WDR 8/7/90, OTH 8/8/90, WDR 8/9/90, OTH 8/9/90, BUF 8/10/90. Jerry Fowler's transcript supplemented these accounts and helped explain some of the legal issues.

pp. 228–230

Spike Jones's transcripts and my memories of traveling with him to Nebraska contributed to this section. See also reporters' accounts of the siting process in other states in OTH 5/27/90 and WDR 8/14/90. The Supreme Court Case was *New York Petitioner v. United States et al.* (No. 91-543); *County of Allegany, New York, Petitioner v. United States et al.* (No. 91-558); *County of Cortland, New York, Petitioner v. United States et al.* (No. 91-563). An article about the case was in NYT 2/10/90.

pp. 230–231

I've juxtaposed issues addressed in Allegany County's Technical Report with the NRC report: Committee to Review New York State's Siting and Methodology Selection for Low-Level Radioactive Waste Disposal of the Board on Radioactive Waste Management Commission on Geosciences, Environment, and Resources National Council, *Review of New York State Low-Level Radioactive Waste Siting Process* (Washington, D.C.: National Academy Press, 1996). All the quotations in this section are from pp. 4-7 in the preface of this report, entitled "Executive Summary." For earlier suggestions from scientists, not living in Allegany County, that the siting process was scientifically flawed, see OTH 1/2/90 and SPEC 4/2/90.

pp. 231–232

The information and quotations came from reporters' accounts in OTH 5/17/90, BUF 5/19/90, WDR 5/24/90, OTH 5/25/90, OTH 5/26/90, OTH 5/30/90, OTH 6/14/90, WDR 6/25/90.

SOURCES

pp. 233–234

The quotations from Thucydides are from Book I, section xxii in *Thucydides: History of the Peloponnesian War, Books I and II*, tr. Charles Forster Smith, vol. 1 (Cambridge: Harvard University Press, 1919; rpt 1980), p. 39.

Appendix

PEOPLE ARRESTED IN ACNAG ACTIONS

Source: *Olean Times Herald* 6/1/89, 1/17/90, 1/19/90, 4/6/90. I am using the people's formal names as reported in the paper. I made only a few corrections when I was sure the person's name was mispelled. The four actions when arrests took place were Belmont 5/31/89, Caneadea 1/16/90, West Almond 1/18/90, Caneadea 4/5/90.

Katherine A. Adlhock, 18, Rochester 4/5/90

Robert J. Albrecht, 45, Alfred Station 1/16/90

Mary E. Allison, 30, Alfred Station 5/31/89

Scott S. Atherton, 23, Middlesex 1/18/90

Charles G. Austin, 65, Belmont 1/18/90

Charles H. Austin, 35, Almond 4/5/90

Dennis L. Aylor, 45, Belfast 1/18/90

Michael J. Babcock, 28, Wellsville 1/16/90

Ermena J. Barber, 65, Friendship 4/5/90

Charles A. Barnes, 46, Fillmore 1/16/90

William H. Bateman, 51, Cuba 1/18/90

Susan Beckhorn, 36, Rexville 4/5/90

Robert D. Blake, 51, Fillmore 4/5/90

Gary M. Brown, 37, Alfred 5/31/89

Joan F. Bryant, 19, Friendship, 4/5/90

Nancy L. Bryant, 42, Friendship 4/5/90

Katherine M. Buffington, 38, Allegany 4/5/90

Carol Burdick, 60, Alfred Station 5/31/89

Susan J. Burlingame, 31, Belmont 5/31/89

Dennis I. Butts, 39, Andover 1/16/90

Sally L. Campbell, 50, Alfred 5/31/89

Stuart L. Campbell, 51, Alfred 5/31/89

Katherine A. Cantwell, 26, Andover 5/31/89

Denny H. Carr, 49, Caneadea 4/5/90

William J. Castle, 48, Belmont 5/31/89

Meredith E. Chilson, 40, Belmont 1/18/90

Carla Coch, 45, Alfred Station 5/31/89

William A. Coch, 40, Alfred Station 5/31/89

Paul J. Curcio, 49, Belfast 1/16/90

Elsie Cushing, 61, Alfred 5/31/89

David Davis, 62, Almond 5/31/89

Jeanette A. Dibrell, 48, Alfred 5/31/89

Sebastian F. Dunne, 29, Hornell 5/31/89

Vicki L. Eaklor, 34, Alfred Station 5/31/89

Christine F. Evans, 44, Almond 4/5/90

Scott C. Feness, 18, Belmont 4/5/90

Kathryn O. Fosegan, 51, Alfred Station 5/31/89

Peter E. Franklin, 34, Rexville 5/31/89

Walter Franklin, 38, Rexville 5/31/89

Shawn P. Frazier, 19, Fillmore 4/5/90

David J. Fredrickson 33, Alfred Station 5/31/89

Jane M. Gaedeke, 31, Perry 5/31/89

Richard K Gaedeke, 63, Dansville 5/31/89

Sharon A. Genaux, 44, Scio 4/5/90

Thomas M. Green, 47, Angelica 4/5/90

William L. Griffin, 61, Tully 4/5/90

Elizabeth M. Groskoph, 50, Wellsville 1/18/90

Ralph G. Groskoph, 53, Wellsville 1/18/90

Cammie J. Gross, 19, Fillmore 4/5/90

Denise L. Hart, 27, Angelica 1/18/90

Anthony R. Harvey, 38, Alfred 5/31/89

Sarah B. Hayden, 18, Lockport 4/5/90

Susan S. Hillman, 35, Cuba 1/16/90

Norman L. Ives, 66, Wellsville 1/16/90

Dale K. Jandrew, 37, Allentown 1/18/90

Michael P. Jaworski, 40, Fillmore 1/18/90

Adam W. Jefferds, 19, Almond 1/18/90

William J. Johnson, 46, Little Genesee 1/18/90

Jeffrey F. Johnston, 42, Alfred 1/18/90

Richard F. Jones, 41, Belfast 5/31/89

Martha Jordan, 44, Alfred 5/31/89

Richard J. Kelley, 37, Almond 5/31/89, 1/18/90

Clarence W. Klingensmith, 74, Alfred Station 4/5/90

Henry J. Koziel, 63, Fillmore 4/5/90

Mark Kurath-Fitzsimmons, 28, Olean 4/5/90

Thomas Lacagnina, 51, Alfred Station 5/31/89

Pamela A. Lakin, 45, Alfred 5/31/89, 4/5/90

Alexandra Landis, 87, Hornell 4/5/90

Gary C. Lloyd, 46, Andover 5/31/89

Jeffrey C. Love, 37, Hartsville 4/5/90

David M. Loveless, 32, Naples 1/18/90

Shirley A. Lyon-Bentley, 57, Wellsville 5/31/89, 4/5/90

Andy L. Mager, 29, Truxton 4/5/90

Frederick G. Marks, 37, Alfred 5/31/89

James M. McCormick, 30, Almond 1/18/90

John D. McQueen, 46, Alfred Station 5/31/89

Shane M. McMahon, 22, Almond 1/18/90

Donald W. Middaugh, 23, Friendship 4/5/90

John D. Nesbitt, 42, Hornell 5/31/89

Anne Marie Oates, 42, Bath 4/5/90

Kevin N. Palmiter, 34, Alfred Station 1/18/90

William D. Parry, 71, Alfred Station 4/5/90

Fleurette M. Pelletier, 57, Angelica 4/5/90

Thomas V. Peterson, 46, Alfred 5/31/89

Ronald G. Preston, 30, Belmont 1/18/90

Dorothea M Rawleigh, 36, Dalton 1/18/90

Andrew M. Robinson, 33, Swain 5/31/89

Challice B. Robinson, 35, Swain 5/31/89

Winnifred A. Robinson, 24, Alfred 5/31/89

Carl J. Root, 28, Allentown 4/5/90

Elizabeth Rossington, 31, Andover 1/18/90

Robert H. Scherzer, 34, Andover 5/31/89

Bradley L. Schiralli, 22, Boliver 4/5/90

Peter R. Schneider, 36, Andover 4/5/90

Thomas J. Setchel, 38, Cuba 4/5/90

E. Jessie Shefrin, 41, Alfred Station 5/31/89

James W. Sheleman, 37, Alfred 5/31/89

Steven L. Skeates, 45, Alfred 5/31/89

Mary Gardner Smith, 31, Wellsville 1/18/90

Eric D. Sommer, 20, Alfred 4/5/90

Linda S. Staiger, 40, Alfred Station 5/31/89

Burton A. Stein, 43, Arkport 1/18/90

Roger B. Van Horn, 41, Alfred Station 5/31/89, 1/18/90

Deborah D. Varney, 38, Belmont 1/18/90

Dennis E. Varney, 38, Belmont 1/18/90

Elizabeth Warek-Fowler, 46, Andover 1/18/90

Floyd W. Wasner, 33, Friendship 4/5/90

Barbara L. Wayson, 52, Caneadea 4/5/90

Roland L. Warren, 74, Andover 4/5/90

Mary Lu Wells, 49, Andover 5/31/89

Charles C. Whitney, 43, Alfred Station 5/31/89

Elspeth A. Whitney 42, Alfred Station 5/31/89

Erich G. Wuersig, 22, Belfast 1/18/90

Klaus Wuersig, 60, Belfast 1/16/90

Lisa R. Wuersig, 19, Belfast 4/5/90

Raymond P. Yelle, 48, Alfred Station 5/31/89, 1/18/90

"Youthful Offender," 17 1/18/90

"Youthful Offender," 17 4/5/90

Index of Names and Identifications

An asterisk indicates transcribed interviews were made with these people. They are available in the archives of the Special Collections, Herrick Library, Alfred University, Alfred, New York and at the County Museum in Belmont, New York. I have included the names of a few people with transcribed interviews who are not mentioned in the text.

I have primarily identified people here in their roles of fighting the nuclear waste dump unless their occupations are important to understanding their roles. Many of these people are identified more completely in the narrative above. I have not included the names of arrestees in this index. They are listed above.